Springer Series in **Materials Science** 29

Edited by U. Gonser

Springer
*Berlin
Heidelberg
New York
Barcelona
Budapest
Hong Kong
London
Milan
Paris
Santa Clara
Singapore
Tokyo*

Springer Series in *Materials Science*

Advisors: M. S. Dresselhaus · H. Kamimura · K. A. Müller
Editors: U. Gonser · R. M. Osgood, Jr. · M. B. Panish · H. Sakaki
Managing Editor: H. K. V. Lotsch

Volume 27 **Physics of New Materials**
Editor: F. E. Fujita

Volume 28 **Laser Ablation**
Principles and Applications
Editor: J. C. Miller

Volume 29 **Elements of Rapid Solidification**
Fundaments and Applications
Editor: M. A. Otooni

Volume 30 **Process Technology for Semiconductor Lasers**
Crystal Growth and Microprocesses
By K. Iga and S. Kinoshita

Volume 31 **Nanostructures and Quantum Effects**
By H. Sakaki and H. Noge

Volumes 1–26 are listed at the end of the book.

Monde A. Otooni (Ed.)

Elements of Rapid Solidification

Fundamentals and Applications

With Contributions by

T. Ando · W. Dmoswki · T. Egami · N.J. Grant · K. Hashimoto
W.L. Johnson · H. Jones · W. Krakow · H. Kronmüller · E.J. Lavernia
H.H. Lieberman · R.C. O'Handley · M.A. Otooni · P. Scharwaechter
C.N.J. Wagner

With 177 Figures

 Springer

Dr. Monde A. Otooni
Department of the Army
U.S. Army Armament Research,
Development and Engineering Center
Picatinny Arsenal, NJ 07806-5000
USA

Series Editors:

Prof. Dr. U. Gonser
Fachbereich 12.1, Gebäude 22/6
Werkstoffwissenschaften
Universität des Saarlandes
D-66041 Saarbrücken, Germany

M.B. Panish, Ph.D.
AT&T Bell Laboratories
600 Mountain Avenue
Murray Hill, NJ 07974-2070, USA

Prof. R.M. Osgood, Jr.
Microelectronics Science Laboratory
Department of Electrical Engineering
Columbia University
Seeley W. Mudd Building
New York, NY 10027, USA

Prof. H. Sakaki
Institute of Industrial Science
University of Tokyo
7-22-1 Roppongi, Minato-ku
Tokyo 106, Japan

Managing Editor:

Dr.-Ing. Helmut K.V. Lotsch
Springer-Verlag, Tiergartenstrasse 17
D-69121 Heidelberg, Germany

ISSN 0933-033X
ISBN 3-540-61791-4 Springer-Verlag Berlin Heidelberg New York

Library of Congress Cataloging-in-Publication Data. Elements of rapid solidification: fundamentals and applications/Monde A. Otooni (ed.). p. cm. – (Springer series in materials science: v. 29) Includes bibliographical references and index. ISBN 3-540-61791-4 (hc: alk. paper) 1. Metals – Rapid solidification processing. I. Otooni, Monde A. II. Series. TS247.E42 1997 669'.94 – dc21 96-53989

This work is subject to copyright. All rights are reserved, whether the whole or part of the material is concerned, specifically the rights of translation, reprinting, reuse of illustrations, recitation, broadcasting, reproduction on microfilm or in any other way, and storage in data banks. Duplication of this publication or parts thereof is permitted only under the provisions of the German Copyright Law of September 9, 1965, in its current version, and permission for use must always be obtained from Springer-Verlag. Violations are liable for prosecution under the German Copyright Law.

© Springer-Verlag Berlin Heidelberg 1998
Printed in Germany

The use of general descriptive names, registered names, trademarks, etc. in this publication does not imply, even in the absence of a specific statement, that such names are exempt from the relevant protective laws and regulations and therefore free for general use.

Cover design: Design & Production GmbH, Heidelberg
Typesetting: Asco Trade Typesetting Ltd., Hong Kong

SPIN: 10545086   54/3144/SPS – 5 4 3 2 1 0 – Printed on acid-free paper

Preface

Almost three and half decades have passed since the discovery of the Rapid Solidification Technology (RST) by *Pol Duwez* and his co-workers at the California Institute of Technology. During this time, a large number of scientists and engineers devoted extensive efforts to the improvement of this technology, characterization of the new products, and the development of new applications of the metastable materials obtained through this technology. As a result of these unprecedented efforts, a new and exciting field of materials science has now emerged with significant potential impact on the materials industry because of the excellent magnetic, mechanical, chemical, and electronic properties of the new materials.

The advent of this technology has opened a noval vista of new scientific challenges to occupy the minds of physicists, chemists, and mathematicians, as well as materials scientists and metallurgists, all working together in quest of a detailed understanding of various phenomena associated with the supercooled state. The results of these research activities are presented in numerous publications in profession-oriented journals, conference proceedings, review articles and books. However, these articles are aimed at the professional researcher rather than college students, while existing textbooks in materials science do not properly introduce young scientists to this new and exciting technology. Consequently, there are gaps in the literature for students who are interested in this field, which this text is designed to bridge.

The objective of this introductory text is to give students a basic understanding of rapid solidification technology and processing. The book is designed for college or first-year graduate students desiring to major in materials science, physics, chemistry, biomedical sciences or other allied fields. It is a product of many prominent scientists working in the field almost from its inception.

The book consists of nine chapters. Chapter 1 provides students with a good understanding of the phenomenological aspects of the field of rapid solidification technology. The chapter deals with such topics as kinetics of nucleation and growth, thermodynamics of supercooled liquid, liquid-to-crystal transition, metallic glasses, and metastable crystalline alloys.

Chapter 2 is primarily devoted to topics of synthesis and processing. It covers such topics as heat transfer and solidification kinetics, techniques of quenching and consolidation processes. The approaches are well explained and several examples are given to further elucidate the concepts.

Chapter 3 is, in the main, devoted to the discussion of structural characterization of rapidly solidified alloys. A large portion of this chapter deals with diffraction studies of amorphous alloys. Additional topics include calculation of the total scattered intensity from amorphous alloys and nanocrystalline solids. Several commonly employed techniques for characterization have also been included.

Chapter 4 presents the basic points of the atomic transport and relaxation phenomenon. In addition, such topics as self-diffusion in amorphous alloys, theory of diffusion in disordered media, diffusion of hydrogen isotopes and light particles in amorphous alloys are all clearly and concisely explained.

Chapter 5 concentrates on mechanical properties and behavior of rapidly solidified alloys. The main topics of this chapter include elastic and anelastic behavior, plastic flow and fracture behavior, strength and hardness, fatigue and wear resistance, and creep and hot deformation behavior.

Chapter 6 presents a thorough account of magnetic and electronic properties of rapidly solidified alloys. It includes topics such as fundamental magnetic properties, magnetic domains, magnetism and short-range order, quasicrystalline materials, and electronic structure. Several interesting applications are also given.

Chapter 7 portrays chemical properties of amorphous alloys. It includes sections on properties of corrosion-resistant amorphous alloys in aqueous solutions, high corrosion resistance of Fe–Cr metalloid alloys, factors determining corrosion resistance, corrosion resistance at high temperature, and the preparation of several electrodes for environmental protection.

Chapter 8 presents several major application areas of rapidly solidified alloys. It gives a good account of the improvement of mechanical properties, size refinements, extended solid solubility, and chemical homogeneity. It also gives several cases of magnetic applications.

Chapter 9 presents a glossary of important terms frequently used in the field of rapid solidification technology.

Many of my colleagues have devoted a major portion of their time and efforts to the preparation of these chapters. I am indeed grateful to all contributors for their timely submissions. These include H. Jones, N.J. Grant, H. Kronmüller, P. Scharwaechter, T. Egami, W.L. Johnson, C.N.J. Wagner, R.C. O'Handley, E.J. Lavernia, H.H. Liebermann, T. Ando, K. Hashimoto, W. Krakow, and W. Dmowski.

Picatinny Arsenal, NJ
July 1997

M.A. Otooni

Contents

1 **Introduction and Background.** By T. Egami and W.L. Johnson
  (With 14 Figures) 1
  1.1 Background 1
  1.2 Liquid-to-Crystal Transition: Undercooling and Nucleation 4
    1.2.1 Thermodynamics and Kinetics of Solidification 4
    1.2.2 Undercooling 5
    1.2.3 Phase Diagram for Metastable States 6
  1.3 Metallic Glasses 8
    1.3.1 Glass Formation by Rapid Quenching 8
    1.3.2 Glass-Forming Composition 8
    1.3.3 Crystallization and Structural Relaxation 10
    1.3.4 Atomic Structure of Metallic Glasses 11
  1.4 Metastable Crystalline Phases 13
    1.4.1 Non-Equilibrium in Crystalline Phases 13
    1.4.2 Two Examples of Solubility Extension. The Ag–Cu
          and Ti–Cu Systems 14
    1.4.3 Metastable Crystalline Phases Not Present
          in Equilibrium – Examples 18
  References 20

2 **Synthesis and Processing.** By N.J. Grant, H. Jones, and E.J. Lavernia
  (With 24 Figures) 23
  2.1 Heat Transfer and Solidification Kinetics 24
  2.2 Droplet Methods 29
  2.3 Spinning Methods 35
  2.4 Surface Melting Technologies 38
  2.5 Consolidation Technologies 42
  References 46

3 **Structure and Characterization of Rapidly Solidified Alloys**
  By C.N.J. Wagner, M.A. Otooni, and W. Krakow (With 32 Figures) . 49
  3.1 Characterization Techniques 49
    3.1.1 Structural Characterization 49
    3.1.2 X-Ray Radial Distribution Function 50

	3.1.3 High-Resolution Electron Microscopy	52
	3.1.4 Differential Scanning Calorimetry – Phase Transformation and Separation	55
	3.1.5 Electrical Resistivity	57
	3.1.6 Microhardness Measurements	61
	3.1.7 Mössbauer Spectroscopy	61
3.2	Total Scattering Intensity from Amorphous and Nanocrystalline Alloys	62
	3.2.1 Atomic Distribution Functions	63
	3.2.2 Scattered Intensity	63
	3.2.3 Reduced Atomic Distribution Functions	64
	3.2.4 Coordination Numbers in Binary Amorphous Alloys	67
	3.2.5 Topological and Chemical Order in Binary Solutions	67
3.3	Diffraction Theory of Powder Pattern Peaks from Nanocrystalline Materials	68
	3.3.1 Fourier Analysis of the Peak Profiles	68
	3.3.2 Integral Breadth of Powder Pattern Peaks	70
3.4	Experimental Diffraction Techniques	72
	3.4.1 Radiation Sources	72
	3.4.2 Diffraction Methods	72
	3.4.3 Variable 2θ Method	72
	3.4.4 Variable λ Method	73
	3.4.5 Analysis of the Diffraction Pattern	73
	a) Total Diffracted Intensity from Amorphous and Nanocrystalline Samples	73
	b) Fourier Analysis of the Profiles of Powder Pattern Peaks	74
3.5	Structure of Amorphous and Nanocrystalline Alloys	74
	3.5.1 Amorphous Beryllium Alloys	75
	3.5.2 Amorphous and Nanocrystalline Vanadium Alloys	77
	3.5.3 Amorphous and Nanocrystalline Tungsten Alloys	79
3.6	Selected Examples of Electron-Microscopy Analysis	84
References		90

4 Atomic Transport and Relaxation in Rapidly Solidified Alloys
By H. Kronmüller and P. Scharwaechter (Appendix) (With 36 Figures)   93

4.1	Basic Equations of Diffusion	94
4.2	Self-Diffusion in Amorphous Alloys	96
	4.2.1 Radiotracer Technique	96
	4.2.2 Non-Equilibrium and Quasi-Equilibrium of Diffusional Properties	98
	4.2.3 Review of Diffusion Data	100
	4.2.4 Diffusion Mechanisms in Amorphous Alloys	102
4.3	Theory of Diffusion in Disordered Media	106
	4.3.1 The Effective-Medium Approximation	106

	4.3.2 Explicite Solutions	106
	4.3.3 The Effective-Medium Approximation for Direct Diffusion Mechanisms	107
	4.3.4 Applications of the "Effective-Medium Approximation"	110
	4.3.5 Molecular Dynamics Simulations and Diffusion Mechanisms	112
4.4	Diffusion of Hydrogen Isotopes and Light Particles in Amorphous Alloys	115
4.5	Magnetic After-Effects and Induced Anisotropies Due to Double-Well Systems in Amorphous Alloys	118
4.6	Viscosity and Internal Friction of Amorphous Alloys	124
	4.6.1 Viscosity Measurements	125
	4.6.2 Internal Friction Measurements	127
Appendix: Microsectioning by Ion-Beam Sputtering – A Powerful Method to Determine Diffusion Profiles		130
References		131

5 Mechanical Properties and Behaviour. By H. Jones and E.J. Lavernia (With 17 Figures) ... 135

5.1	Elastic and Anelastic Behaviour	135
5.2	Plastic Flow and Fracture Behaviour	137
5.3	Strength and Hardness	138
5.4	Fatigue and Wear Behaviour	142
5.5	Creep and Hot Deformation Behaviour	146
References		151

6 Magnetic and Electronic Properties of Rapidly Quenched Materials
By R.C. O'Handley and H.H. Liebermann (With 23 Figures) 153

6.1	Rapidly Quenched Alloys	155
	6.1.1 Amorphous Alloys	155
	6.1.2 Nanocrystalline Alloys	156
6.2	Fundamental Magnetic Properties	157
	6.2.1 Magnetic Moments and Curie Temperatures	157
	6.2.2 Magnetic Anisotropy	160
	6.2.3 Magnetostriction	161
6.3	Domains and Technical Properties of Amorphous Alloys	163
	6.3.1 Domains	163
	6.3.2 Coercivity	165
	6.3.3 Magnetic Hardening	166
	6.3.4 Induced Anisotropy	167
6.4	Magnetism and Short-Range Order	169
	6.4.1 Ingredients of Short-Range Order	169
	6.4.2 Random Local Anisotropy	170

6.5 Electronic Structure of Amorphous Alloys ... 172
   6.5.1 Chemical Bonding ... 172
   6.5.2 Split d Bands and p-d Bonding ... 173
   6.5.3 Electron Transport ... 177
6.6 Applications ... 179
   6.6.1 Distribution Transformers ... 182
   6.6.2 Electronic Article Surveillance Sensors ... 182
   6.6.3 Magnetic Recording Media ... 183
   6.6.4 Permanent Magnets ... 184
6.7 Conclusion, Outlook ... 184
References ... 185

7 Chemical Properties of Amorphous Alloys. By K. Hashimoto (With 22 Figures) ... 187

7.1 Corrosion-Resistant Alloys in Aqueous Solutions ... 187
   7.1.1 High Corrosion Resistance of Amorphous Fe–Cr Alloys ... 187
   7.1.2 Factors Determining the High Corrosion Resistance of Amorphous Alloys ... 188
      a) High Activity of Amorphous Alloys ... 188
      b) Homogeneous Nature of Amorphous Alloys ... 191
      c) Beneficial Effect of Phosphorus in Amorphous Alloys ... 192
   7.1.3 Recent Efforts in Tailoring Corrosion-Resistant Alloys ... 195
      a) Aluminum-Refractory Metal Alloys ... 195
      b) Chromium-Refractory Metal Alloys ... 196
7.2 Corrosion-Resistant Alloys at High Temperatures ... 200
7.3 Electrodes for Electrolysis of Aqueous Solutions ... 202
   7.3.1 Electrode Materials ... 202
   7.3.2 Preparation of Electrodes ... 205
7.4 Catalysts for Prevention of the Greenhouse Effect and Saving the Ozone Layer ... 209
   7.4.1 CO_2 Recycling ... 210
   7.4.2 Catalysts for the Decomposition of NO_x ... 213
   7.4.3 Catalysts for the Decomposition of Chlorofluorocarbons ... 214
7.5 Concluding Remarks ... 214
References ... 215

8 Selected Examples of Applications. By H.H. Liebermann, N.J. Grant, and T. Ando (With 8 Figures) ... 217

8.1 Improvement of Mechanical Properties ... 218
   8.1.1 Size Refinement ... 218
   8.1.2 Extended Solid Solubility ... 222
   8.1.3 Chemical Homogeneity ... 224
8.2. Magnetic Applications ... 225
   8.2.1 Magnetic Properties and Applications ... 226

         8.2.2 Power Magnetic Applications 227
         8.2.3 Specialty Magnetic Applications 227
    8.3 Joining Applications 229
    8.4 Current Limitations and Future Directions 231
    Further Reading 235

9 **Glossary of Important Terms.** By T. Egami, M.A. Otooni,
    and W. Dmoswki 237

Subject Index 243

List of Contributors

Ando, T., Department of Mechanical, Industrial and Manufacturing Engineering, Northeastern University, 334 Snell Engineering Center, Boston, MA 02115, USA

Dmoswki, W., Department of Materials Science and Engineering, University of Pennsylvania, LRSM, 3231 Walnut Street, Philadelphia, PA 19104-6272, USA

Egami, T., Department of Materials Science and Engineering, University of Pennsylvania, LRSM, 3231 Walnut Street, Philadelphia, PA 19104-6272, USA

Grant, N.J., Department MS & E, Rm 8-413 / MIT, Cambridge, MA 02139, USA

Hashimoto, K., Institute for Materials Research, Tuhoku University, 2-1-1 Katahira, Aoba-ku, Sendai, 980-77 Japan

Johnson, W.L., California Institute of Technology, 138-78 Keck Laboratory, Pasadena, CA 91125, USA

Jones, H., Department of Engineering Materials, The University of Sheffield, Sir Robert Hadfield Building, Mappin Street, Sheffield S1 3JD, UK

Krakow, W., Mountainside Trail, Peekskill, NY 10566, USA

Kronmüller, H., Max-Planck-Institut für Metallforschung, Heisenbergstrasse 1, D-70569 Stuttgart, Germany

Lavernia, E.J., Department MS & E, University of California at Irvine, Irvine, CA 92717, USA

Liebermann, H.H., AlliedSignal Inc., Amorphous Metals, 6 Eastman Road, Persippany, NJ 07054, USA

O'Handley, R.C., Department of Materials Science and Engineering, Massachusetts Institute of Technology, Rm 4-405, Cambridge, MA 02139, USA

Otooni, M.A., US Army ARDEC, Rm-45 B-355, Picatinny Arsenal,
NJ 07806-5000, USA

Scharwaechter, P., Institut für Theoretische und Angewandte Physik,
Universität Stuttgart, Pfaffenwaldring 57, D-70569 Stuttgart, Germany

Wagner, C.N.J., University of California at Los Angeles, 6531 Boetler Hall,
Los Angeles, CA 90024, USA

1 Introduction and Background

T. Egami and W.L. Johnson

The modern science and technology of rapid solidification has its beginnings about 35 years ago with the work of Pol Duwez and his coworkers at the California Institute of Technology. They developed methods for chilling a thin layer of liquid metal by rapid heat transfer to a highly conductive and relatively cold substrate. Despite its relatively recent emergence, Rapid Solidification Technology (RST) has had a substantial impact on our fundamental understanding of materials synthesis by solidification as well as on our ability to develop materials for technological applications. Rapid solidification can be somewhat arbitrarily defined as any solidification process in which the rate of change of temperature (the cooling rate) is at least 10^2 K/s and typically of the order of 10^6 K/s. In fact, cooling rates as high as $10^{10}-10^{11}$ K/s are achieved in laboratory experiments where melting is induced using a pulsed laser or other intense energy source. RST is the science and practice of melt solidification at such high rates of cooling.

1.1 Background

The effects of RST on phase equilibria and microstructure of materials are readily observed. They include: (*i*) the reduction of grain size and the typical scale of microstructure as cooling rate increases; (*ii*) increased chemical homogeneity with increasing cooling rate; (*iii*) extension of solubility and homogeneity ranges of equilibrium phases with increasing cooling rate; (*iv*) production of metastable crystalline phases not present in equilibrium; (*v*) the failure of the liquid to undergo crystallization entirely, resulting in the formation of a non equilibrium glassy phase.

Glassy phases are, in fact, a new class of metallic materials which were, prior to 1960, not found to exist under conventional processing conditions. All of these effects will be described in more detail in this chapter. In the remainder of the present section, we will briefly outline the principles which underlie the techniques itself and also describe some of the experimental implementations of the technique.

The practice of rapid solidification involves the removal of heat from a molten sample at high rates. Generally, this is accomplished through conductive heat transport, sometimes acting in concert with convective heat transport. Radiative heat transport is for the most part too inefficient to permit the achievement of very rapid cooling. For conductive heat transport, the time evolution of the sample's temperature distribution is governed by the Fourier heat flow equation

$$dT(x,t)/dt = -\kappa \nabla^2 T(x,t), \tag{1.1}$$

where $T(x,t)$ represents the spatial and temporal distribution of temperature, and κ is the thermal diffusivity of the sample and has the dimensions of cm²/s. The thermal conductivity, K, of the sample is given by $K = \kappa/c_p$, where c_p is the specific heat of the material (in units of erg/cm³·C) at constant pressure. Together with suitable boundary conditions, this differential equation describes the time evolution of the sample's temperature profile. The general solutions to this differential equation are discussed in many other textbooks. Here we point out only some of the main features relevant to rapid solidification.

The solutions to the Fourier heat flow equation can be described as functions of the dimensionless variable $y = \kappa t/x^2$, where x is the spatial coordinate and t is the time. For instance, when a thin layer (thickenss L) of a liquid metal at a high temperature T_i (initial temperature) comes into contact with a massive cold substrate at temperature T_f (final temperature), the characteristic solution to the Fourier equation can be written as a universal function of the dimensionless variable y. Practically speaking, this means that the evolution of the temperature in the liquid layer can be described as a dimensionless function of $y = \kappa t/L^2$, where κ is the thermal conductivity of either the liquid (if the substrate is a relatively poor conductor) or the thermal conductivity of the substrate (for a substrate of relatively high thermal conductivity). Generally speaking, the solution to the heat flow equation involves both the thermal conductivity of the substrate and the liquid. If the temperature difference is $\Delta T = T_f - T_i$, then the typical cooling rate in the experiment is given by

$$dT/dt = \frac{\Delta T \kappa}{y_c L^2}, \tag{1.2}$$

where y_c is a characteristic value of the dimensionless variable y. Typically, y_c will have a value on the order of unity.

Equation (1.2) allows one to estimate the characteristic cooling rate in a real rapid solidification problem. For rapid quenching of a thin liquid layer on a highly conductive copper substrate with $\kappa \approx 0.3$ cm²/s, $y_c \approx 3$, and $L = 100\,\mu\text{m} = 0.01$ cm, and $\Delta T = 1000\,°\text{C}$, one obtains a cooling rate of $10^6\,°\text{C/s}$. This is a typical cooling rate obtained in rapid solidification by "splat quenching" or melt spinning. In splat quenching, as originally practiced by *Duwez* and colleagues, a liquid drop is rapidly flattened between two rapidly closing copper plates. The typical final thickness of the "splat" is $\approx 50\,\mu\text{m}$. In melt spinning, a thin layer of liquid is ejected through a slotted nozzle onto a rotating copper drum. The liquid is pulled over the drum surface (by wetting), cools, and solidifies while in

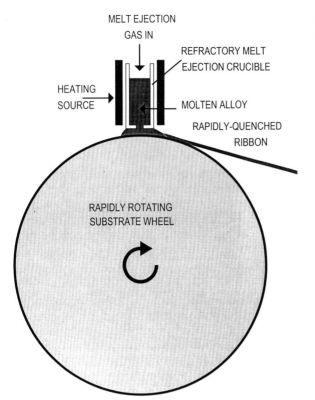

Fig. 1.1. Schematic of melt spinning method

intimate contact with the copper drum. The thickness of the solidified layer is about 20–50 µm. Figure 1.1 illustrates the melt spinning process. Once again, the critical cooling rate obtained in melt spinning is of the order of 10^6 °C/s. The exact cooling rate depends on many details such as the degree of intimate contact between the melt and the rotating wheel, the thermal conductivity of both the liquid and copper wheel, etc.

Another type of rapid solidification process frequently used commercially is called spray atomization. Here, the liquid is atomized into rapidly moving droplets by a gas jet and sprayed into a gaseous cooling medium such as helium or argon. In this process, the thermal diffusivity of the gas becomes the rate limiting factor in determining the cooling rate when the velocity of the droplet is relatively small. For an atomized droplet diameter of 100 µm, and helium cooling gas, one might typically obtain a cooing rate of 1000 °C/s.

There are a variety of other variations of the rapid solidification process. For a review of these, the reader is referred to several good reviews [1.1, 2].

1.2 Liquid-to-Crystal Transition: Undercooling and Nucleation

When a liquid freezes into a crystalline solid, the transformation does not happen all of a sudden. Freezing starts somewhere, usually at the edge or the surface of the liquid, and gradually spreads to other parts. This is partly because the temperature in the liquid is usually inhomogeneous, but also because the transition from the liquid state to the crystalline state is a discontinuous process, involving nucleation and growth [1.3]. The crystalline solid phase and the liquid phase are fundamentally different in their atomic structure, one having long range periodicity and the other being aperiodic and amorphous. Consequently, the liquid–crystalline solid interface is well defined, having the thickness of a few atomic distances, and the free energy of the interfacial region is higher than those of either the solid or the liquid. Therefore, in order to create a nucleus of a solid in the liquid, an energy cost must be paid because the solid–liquid interface has to be created. This energy cost represents a barrier for nucleation, and can only be overcome by thermal activation. But once this nucleation barrier is overcome the solid can grow by the advancement of the solid-liquid interface. This process is also thermally activated, involving diffusion.

1.2.1 Thermodynamics and Kinetics of Solidification

A simple way to estimate the magnitude and temperature dependence of the barrier height is to consider how much the free energy changes when a small spherical crystalline solid nucleus is created. The change in the free energy is given by

$$\Delta G(r) = -\tfrac{4}{3}\pi r^3 \Delta G_v + 4\pi r^2 \gamma, \tag{1.3}$$

where ΔG_v is the gain in the free energy per unit volume due to a liquid transforming into a solid, r is the radius of the nucleus, and γ is the energy per unit area of the interface between the liquid and the solid. As shown in Fig. 1.2 this

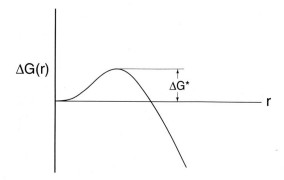

Fig. 1.2. Gibbs' free energy of a crystalline nucleus in a liquid as a function of the radius of the nucleus

energy change is initially positive, goes through a maximum, and only after that becomes negative. This means that the system eventually gains energy when the nucleus is large enough; but initially there is an energy barrier. The barrier height is given by

$$\Delta G^* = \frac{16\pi\gamma^3}{3(\Delta G_v)^2}. \tag{1.4}$$

This is the energy barrier which must be overcome in order to initiate the solidification process via nucleation.

1.2.2 Undercooling

At the melting temperature (T_m), ΔG_v is zero, since at T_m the liquid and solid phases are at equilibrium and the free energies of these two phases are equal. This means that the energy barrier ΔG^* is infinite, and therefore freezing would *never occur at T_m*. Below the melting temperature, ΔG_v is proportional to $\Delta T = T_m - T$. Thus, ΔG^* is inversely proportional to $(T_m - T)^2$, and is still quite large unless T is substantially below T_m. Therefore, a liquid can easily be undercooled or brought below T_m without solidification for a certain amount of time.

The time necessary for a nucleus to be formed, or the nucleation time, decreases with increasing undercooling, since ΔG^* is decreased as T is lowered. But as the temperature is further reduced atomic mobility through diffusion becomes less, and the time for the nucleus to grow becomes longer. Therefore, the amount of time for a liquid at a temperature T to become a solid, t_s, is a highly nonlinear function of T. When T is just below T_m, t_s decreases rapidly with decreasing temperature, but at lower temperatures it becomes longer again. Figure 1.3 is the plot of t_s on a logarithmic scale against temperature which is on a linear scale. This type of plot is known as the *T-T-T* (*Time-Temperature-*

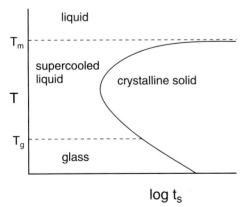

Fig. 1.3. T-T-T (Time-Temperature-Transformation) diagram for liquid to crystal transformation. The horizontal axis is the time a liquid takes to transform itself into a crystal, in a logarithmic scale. By circumventing the nose of the C-curve by rapid cooling a glass can be obtained

Transformation) diagram. The locus of t_s is almost always a C-shaped curve, for the reason discussed above.

1.2.3 Phase Diagram for Metastable States

The phase diagrams we normally use describe the state of matter in thermal equilibrium. However, during rapid cooling the material often does not have time to reach the equilibrium state, and can remain in a metastable state. A glass indeed is an example of such a state. In order to describe such states we need a phase diagram for metastable states. The degree of metastability, however, is process dependent and varies greatly from one case to another. To avoid this ambiguity it is useful to consider an intermediate, somewhat artificial, state of matter which evolves with a time scale longer than the thermal diffusion time, but shorter than the atomic diffusion time. In this system temperature is homogeneous, but atomic diffusion is totally suppressed.

As an example let us consider the left portion of a binary phase diagram given as Fig. 1.4 in which the solid solution (α-phase) is in equilibrium with a liquid alloy. The free energy diagram of this system at a temperature T' is given in Fig. 1.5. If we consider the fully equilibrium states, the tie-line in the free energy diagram defines the solidus (x_s) and liquidus (x_L) compositions. If diffusion is suppressed, however, the phase separation into a solid with the composition x_s and a liquid with the composition x_L is not allowed. In this case for the concentration greater than x_0, the liquid is the most stable phase since it has the lowest free energy. The locus of x_0 at various temperatures is called the T_0 line [1.4]. The T_0 line defines the freezing or melting temperature of a homogeneous solid solution when phase separation is totally suppressed.

On the right-hand side of Fig. 1.5, the free energy of a metastable phase (θ-phase) is lower than those of a liquid and a solid solution (β-phase) in the

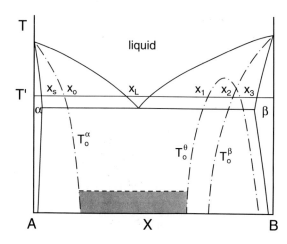

Fig. 1.4. A binary phase diagram showing the T_0 line (chained curves). The α-phase is a solid solution of B atoms in the A matrix, and the β-phase is a solid solution of A in the B matrix. The θ-phase is a metastable phase. Hatched range indicates the glass forming composition

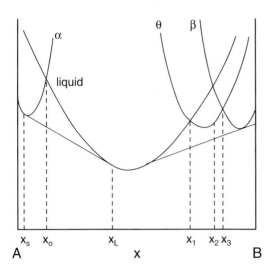

Fig. 1.5. Free energy diagram of a binary system of which the phase diagram is shown in Fig. 1.3. The figure shows the free energy G of various phases (α, liquid, β, θ) vs composition x at temperature T'. Between x_0 and x_1, liquid (glass) is the most stable phase, except for the mixture of the α and β phases. Between x_1 and x_3, the θ-phase is most stable if phase separation is suppressed. The loci of x_0, x_1, x_2 and x_3 at various temperatures define the T_0^α, T_0^θ and T_0^β lines

composition range between x_1 and x_3, although it is higher than the tie-line connecting the free energies of the liquid and the β-phase. Thus, even though the θ-phase is metastable and does not show up in the equilibrium phase diagram, its T_0 line is above that of the β-phase over a certain composition range. When a liquid of this composition is rapidly cooled the product is likely the metastable θ-phase rather than the stable β-phase.

A metastable phase can also be obtained by taking advantage of the kinetics. In Fig. 1.6, the T-T-T diagrams for a slow-forming stable phase and a fast-forming metastable phase are exhibited. Shown is also the history of one particular cooling procedure, or the temperature of the liquid as a function of time,

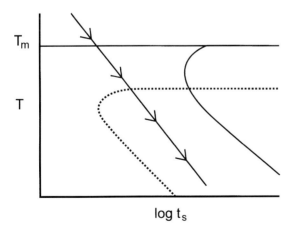

Fig. 1.6. The T-T-T diagram for a slow forming stable phase (*solid line*) and a fast forming metastable phase (*dashed line*)

plotted by a curve with arrows. In this particular case, the curve does not touch the C-curve of the stable phase, but it impinges upon the C-curve of the metastable phase, thus producing the metastable phase. This subject will be discussed in further detail in Sect. 1.4.

1.3 Metallic Glasses

1.3.1 Glass Formation by Rapid Quenching

If a molten liquid alloy is cooled rapidly enough, it is possible to avoid touching the "nose" of the C-curve of any phase in the T-T-T diagram. In this case, the nucleation of crystalline particles is suppressed altogether, and the liquid remains supercooled. As the temperature of the liquid is lowered its viscosity increases very quickly, by many orders of magnitude, and below a certain temperature, the liquid is so viscous that it behaves like a solid. This solid obtained by freezing the liquid is a glass, and the temperature below which the liquid behaves like a solid is called the glass transition temperature, T_g. A metallic glass of Au–Si was first obtained by rapid quenching of the melt by *Duwez* and his students in 1960 [1.5]. They then discovered that various alloys of transition metals and metalloids can be quenched into a glass.

1.3.2 Glass-Forming Composition

It is interesting to note that the principle of rapid cooling discussed before is not limited to specific materials, but applies in general. Therefore, if one could cool the liquid fast enough, *any material can be made a glass*. If the material can actually be made a glass or not is determined by whether the critical cooling rate for glass formation can be experimentally attained or not. *Turnbull* was the first to recognize this intriguing fact and established the theoretical basis for the existence of a metallic glass [1.6].

The critical cooling rate necessary to form a glass is determined by the position of the "nose" of the C-curve in the T-T-T diagram, and varies greatly from one system to another. For a pure metal, the critical rate is estimated to be over 10^{12} K/s, while for many oxide systems it is only about 1–10 K/s. In comparison, the cooling rate attainable with the rapid cooling apparatus such as the one used by Duwez's group or the ones used in industries producing metallic glasses is of the order of 10^6 K/s. Thus, only certain alloy systems over a certain composition range can be quenched into a glass.

The position of the C-curve affects also the stability of the glass once it is formed. If the C-curve is very much to the left, the crystallization time t_s would be short even at temperatures well below T_g so that the glass will be highly metastable and eventually crystallize after a while. Pure or nearly pure metals may be made a glass by vapor deposition at very low temperatures, but they crystallize before they are brought up to room temperature. On the other hand, if the C-curve is very much to the right, t_s will be so far below T_g that the glass practically is a stable state of matter. This is why ancient glasswares remain glassy today, even though glasses are metastable.

The position of the nose of the C-curve, and thus, the glass forming ability, depends critically upon composition. It is not uncommon that the position of the C-curve shifts by orders of magnitude by small changes in composition. The composition dependence of the relative stability of a glassy phase may be best illustrated by the T_0 lines in the phase diagram [1.7]. If the T_0 line comes down below T_g, as shown in Fig. 1.4, for a composition beyond the T_0 line, the glass is the stable phase as long as phase separation is suppressed. Beyond the T_0 line crystallization occurs only after phase separation, and therefore, it is a slow process. The C-curve of such a process is shifted much to the right, and it is easy to circumvent the nose of this C-curve by rapid cooling. Thus, the glass forming composition can roughly be defined by the T_0 line falling below T_g. For this reason, the glass forming composition range is often characterized by the eutectic phase diagram. Near the eutectic composition, T_0 is low and the liquid is relatively more stable; thus, glasses are more easily formed.

The position of the T_0 line, and therefore the glass formability, is influenced by various chemical and physical factors. Among them, the size difference among the constituent atoms seems to be the most dominant factor [1.7]. In the crystalline state substituting a smaller atom with a larger atom will expand the lattice. If the concentration of the substituted element is high enough, the lattice will become unstable, and such a crystal will not form during solidification. Thus, alloying various elements with different atomic sizes destabilizes the crystalline phase and shifts the C-curve to the right. The principal role of alloying is *not* to stabilize the liquid state by chemical interaction, but to destabilize the crystalline state by the size effect.

Today, a large number of alloy systems are known to form stable glasses [1.8]. Typical glass forming compositions are transition metal–metalloid alloys such as $TM_{100-x}M_x$, where TM is an alloy of Fe, Ni, Co and Cr, M is a metalloid element such as B, Si, C and P or their mixture, and roughly $15 < x < 25$, transition metal alloys such as $Zr_{100-x}Cu_x$, where $40 < x < 75$, and aluminum–transition-metal alloys such as $Al_{100-x}Fe_yCe_{x-y}$, where $9 < x < 15$. The exact range of glass-forming composition depends on the details of the production method; but since the critical cooling rate very strongly depends upon composition, the glass forming ranges determined by various researchers are remarkably similar. Figure 1.7 depicts one of such composition ranges for the Fe–B–Si alloy system [1.9]. Glass forming is easiest at the center of the range shown. The glass transition temperature for these compositions weakly depends on the composition, and is in the range of about 300–450 °C. (Metallic glasses exhibit unique properties as discussed in this volume and other literature [1.1, 2, 10]. Many of these properties are of great interest for application, and metallic glasses are becoming one of the most important advanced materials today.)

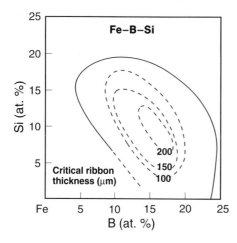

Fig. 1.7. Glass forming range and crystallization temperature for Fe–B–Si alloys (inside the solid curve, after Waseda et al.). Numbers denote the maximum thickness (in μm) of the ribbon which can be cast as metallic glass

The ease of glass formation appears to increase with the increasing degree of complexity of the composition. This is reasonable since it would be more difficult for a complex system to crystallize into an ordered structure. Recently, some quinternary systems such as $(Ti, Zr)_{55}(Ni, Cu)_{25}Be_{20}$ [1.11] were found to form a glass so easily that rapid cooling is no longer necessary. This removed the necessity that one of the dimensions of the metallic glasses has to be small in order to extract heat rapidly during quenching, which restricted the metallic glasses to the form of ribbons or sheets with the thickness of 20 to 50 μm. They now can be cast in a bulk form and can be formed or machined into various useful shapes [1.12, 13].

1.3.3 Crystallization and Structural Relaxation

A glass is inherently metastable or unstable and eventually crystallizes along the *C-curve* shown in Fig. 1.3. Since crystallization is a kinetic phenomenon, if a glass is heated up with a constant heating rate, the temperature at which the glass crystallizes, T_x, depends upon the heating rate. For a constant heating rate, the temperature dependence of T_x is Arrhenian, with the apparent activation energy ranging typically from 2 to 5 eV. The product of crystallization need not be a stable phase. As shown in Fig. 1.6, often a metastable phase is observed before the stable phase appears. This provides an interesting method to obtain a desired metastable phase via a glassy state.

When a metallic glass is annealed, or held at an elevated temperature for a prolonged time, marked variations in a number of properties are observed before crystallization sets in. This process is called structural relaxation [1.14]. During the process of rapidly cooling a liquid alloy to form a glass at a temperature somewhat above T_g, the viscosity of the liquid becomes so large that the structure of the liquid would no longer be able to follow the changing temperature and becomes virtually frozen. A glass thus produced retains the structure at that temperature, called the fictive temperature, which depends upon the cooling rate. The higher the cooling rate, the higher is the fictive temperature. When a glass with a high fictive temperature is subsequently annealed, the structure gradually reverts itself to a more stable structure and its fictive temperature approaches the annealing temperature. This structural relaxation occurs even well below the glass transition temperature, and is accompanied by various changes in the properties including magnetic as well as mechanical properties. Incidentally, the structural relaxation is *not an incipient crystallization*. It results in a more stable glass which is structurally totally distinct from the crystalline state.

It is useful to recognize the distinction between Topological Short Range Ordering (TSRO) and Compositional Short Range Ordering (CSRO) during structural relaxation [1.15]. The topology of the structure, or the way atoms are bonded together, described by its fictive temperature cannot attain equilibrium except for the immediate vicinity of T_g. Thus, the relaxation of the topology of the structure is largely irreversible and the effects of additional annealing treatments are always additive. However, the compositional short-range order among similar atoms, for instance among the transition-metal elements, can reach an equilibrium in a relatively short time, and can exhibit a reversible behavior. Many magnetic properties are more sensitive to CSRO than to TSRO and thus exhibit reversible annealing behavior [1.14, 15].

1.3.4 Atomic Structure of Metallic Glasses

Diffraction measurements of X-rays, electrons and neutrons are commonly used in characterizing the atomic structure of crystalline solids. Diffraction patterns thus obtained from crystalline solids show sharp peaks (Bragg peaks) or spots (Laue spots) corresponding to the periodicity and symmetry of the crystalline lattice. Diffraction patterns from a glass, on the other hand, show no sharp features. If one places a metallic glass sample in a powder diffractometer, the diffraction intensity pattern will consist of several broad peaks as shown in Fig. 1.8. Here, the diffracted X-ray intensity is plotted against the diffraction vector Q ($Q = 4\pi \sin\theta/\lambda$, where θ is the diffraction angle and λ is the wavelength of the X-ray).

These broad peaks in the diffraction pattern cannot be interpreted in the same manner as in the crystallographic analysis of the crystalline structure. It immediately tells us that the structure of a glass is not periodic, and a wide distribution of length scales is involved. A carefully measured diffraction pattern can be reduced to the atomic interference function, and then to the atomic Pair-Distribution Function (PDF) by the Fourier transformation [1.16], as shown in Fig. 1.9. The PDF, $\rho(r)$, describes the probability of finding an atom at a distance r, centering on another atom. The PDF of a metallic glass is characterized by a relatively sharp first peak which corresponds to the distance to the nearest-neighbor atoms, and oscillations which become weak with the

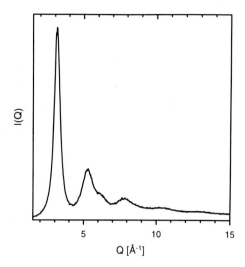

Fig. 1.8. Intensity of X-rays scattered from a metallic glass, as a function of diffraction vector Q

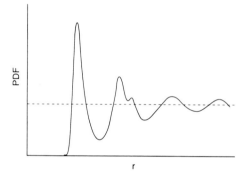

Fig. 1.9. Atomic pair-distribution functions of a metallic glass, obtained by Fourier-transforming the normalized X-ray or neutron scattering intensity

increased distance. By integrating the PDF times $4\pi r^2$ over the first peak, one obtains the number of nearest neighbors, which is usually about 13.

However, the PDF gives only one-dimensional information regarding the atomic distances, and does not permit the actual three-dimensional structure to be known directly. Thus, it is necessary to build a three-dimensional model to further interpret the PDF. The first successful models of the liquid structure were proposed simultaneously by *Bernal* and *Scott* in 1960 [1.17, 18], and were named the dense random packing model. The physical model made by Bernal consisted of many steel balls randomly packed in a bag, glued together with wax. By analyzing this model, Bernal noticed that local clusters rarely found in crystals such as an icosahedron (Fig. 1.10) are abundantly found in this model. Since an icosahedron is characterized by *five-fold* symmetry, it is not compatible with crystalline periodicity, except in complex multi-element compounds. Earlier in 1952, *Frank* [1.19] has already pointed out that in isolation an icosahedron is the most stable form of a cluster made of 13 atoms and that the presence of such clusters stabilizes the liquid structure.

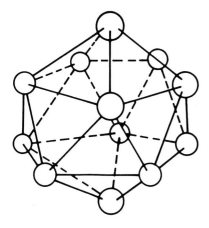

Fig. 1.10. Icosahedral cluster of atoms

In liquids and glasses, by removing the constraint of periodicity, local structures which are usually not allowed in a crystal but are locally more stable can be found. A glass is formed with the principle of minimizing the local energy at the expense of global energy, while a crystal is formed by globally minimizing the total energy at an occasional expense of the local energy. Thus, a glass is formed when the kinetic constraint does not allow a wide configurational space to be probed by the system. On the other hand, when the atomic mobility is high and the system is allowed to look for the free energy minimum in the entire phase space usually a crystalline state is achieved. A solid with the overall icosahedral symmetry was later discovered [1.20], and was named a quasicrystal. However, the quasicrystal is in a completely ordered state with two periodicities which are incommensurate to each other, and its structure is not directly related to that of a glassy state which is highly disordered with no periodicity [1.21].

The recent discovery of glass compositions with very low critical cooling rate for glass formation is likely to promote the application of metallic glasses even further. For these compositions a bulk glass can be cast easily and inexpensively. By machining these bulk glasses complex tools such as gears can be produced. It is most likely that they will find major markets in the near future. These applications are discussed in more detail later in this book.

1.4 Metastable Crystalline Phases

1.4.1 Non-Equilibrium in Crystalline Phases

The principles that govern the formation of a metastable crystalline phase during undercooling from the melts are much the same as those which govern the formation of metallic glasses. Like the glassy phase, metastable crystalline phases can be characterized by an excess free energy and enthalpy. Upon transformation to the equilibrium phase, the excess enthalpy is released as stored heat and the excess free energy is given up. To produce metastable crystalline phases, one must suppress the attainment of full equilibrium during solidification from the melt.

A solidified crystalline alloy can be out of equilibrium in at least two distinctly different ways:

1) Non-equilibrium with respect to the phases present;
2) Non-equilibrium with respect to the chemical composition of a phase.

In the first case, a phase is present which does not exist in the equilibrium phase diagram, i.e., the alloy has a non-equilibrium crystal structure. In the second case, the phases present belong to the equilibrium diagram, but the compositions of the phases lie outside of the equilibrium composition fields. The second case is often referred to as "solubility extension" or "extension of the equilibrium homogeneity range". Rapid quenching of a liquid limits both the time available for nucleation of phases (often favoring nucleation of a non-equilibrium phase) and also restricts chemical diffusion and thereby leads to deviations from chemical equilibrium. We will discuss these two cases separately in the following two sections. No attempt will be made to review the many observations of metastable phases obtained by rapid quenching in the literature. Rather, the underlying principles will be discussed using selected examples.

1.4.2. Two Examples of Solubility Extension. The Ag–Cu and Ti–Cu Systems

The attainment of full chemical equilibrium in a solidified sample requires redistribution of elements within and among the phases present to achieve a constant chemical potential throughout the system for each of the components present. For each phase, this implies the absence of composition gradients within the phase. For two (or more) phases, the compositions of the phases must be chosen to satisfy the common tangent rule. This guarantees a common chemical potential for both (or more) components in the two phases. During rapid solidification, the required chemical redistribution may be kinetically suppressed. To illustrate the underlying principles, we choose two examples.

The phase diagram of the Ag–Cu system is shown in Fig. 1.11. The T_0 line for the crystalline fcc phase (Sect. 1.2) is also illustrated. In equilibrium at ambient temperature, one has two fcc-phases which are, respectively, Cu-rich and Ag-rich. During the earliest rapid solidification experiments, Duwez [1.22] showed that rapid quenching the Ag–Cu melt at cooling rates of 10^6 K/s produces a continuous series of single-phase fcc alloys over the entire range of compositions. This is referred to as complete solubility extension. To understand this behavior, consider the corresponding free energy diagram in Fig. 1.12. For an alloy of the eutectic composition at modest undercooling (above the T_0 curve but below the eutectic) there is no driving force to form an fcc phase of

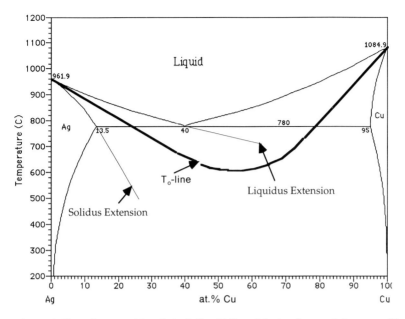

Fig. 1.11. Phase diagram of Ag–Cu including T_0 line of the fcc phase and the metastable extension of the solidus and liquidus curves of the terminal fcc phases to temperatures below the eutectic

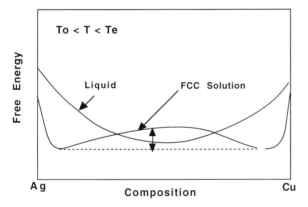

Fig. 1.12. Free energy diagram of the Ag–Cu system showing the absence of a driving force for polymorphic crystallization of the melt below the eutectic temperature, the dashed line represents equilibrium at the temperature just below the eutectic

the eutectic composition. Crystal phase formation at this temperature requires a crystal of composition differing from that of the eutectic liquid. Due to the small undercooling, the nucleation barrier for such a crystal will be large as discussed in Sect. 1.2. When the time available is restricted by rapid cooling, such nucleation may not occur. At larger undercooling, below the T_0 line, the nucleation barrier is reduced (Sect. 1.2). Further, a driving force exists for partitionless crystallization undercooled with respect to the T_0 line. The driving force for partitionless crystallization is smaller than that with a composition shift, but the latter requires that the liquid undergoes a finite composition fluctuation which in turn requires atomic diffusion in the liquid. During rapid solidification at 10^6 K/s, the time available for such a composition fluctuation in the liquid state is severely limited; typically the time available during cooling at 10^6 K/s is less than 1 ms. Crystal nucleation without composition shift now becomes favored with progressively greater undercooling. At sufficiently high cooling rates and correspondingly deeper undercooling, the driving force for both partitionless or non-partitionless crystallization becomes sufficient to allow nucleation, but partitionless crystallization becomes the favored kinetic path since it does not involve composition shifts and the required atomic diffusion in the undercooled melt.

One might now ask way the composition of the growing crystal (once nucleated) remains the same as that of the remaining liquid which surrounds it. In fact, the common tangent rule for chemical equilibrium would suggest that a growing crystal at a given undercooling having the composition of the extended solidus (Fig. 1.11) could only be in chemical equilibrium with a liquid having the liquidus composition at the same undercooling. Our growing crystal and remaining liquid should develop a composition difference. The growing crystal should approach this composition as solidification proceeds. That this does not occur during rapid solidification can also be explained in terms of kinetic constraints. When a solidification front (our crystal–liquid interface) advances with sufficient velocity, insufficient time is available to develop a composition gradient within the liquid ahead of the moving interface. Again, the development of such a gradient requires atomic diffusion. Such diffusion is kinetically suppressed by the rapid advance of the solidification front. This phenomenon is called "solute trapping". At sufficiently high solidification velocity, the solidifying crystal will be kinetically constrained to have the same composition of the parent liquid. Partitioning along the liquid–solid interface is suppressed at high solidification velocities. A detailed theoretical analysis of solute trapping can be found in [1.23]. Here, the deviation of the crystal composition from its equilibrium value for a terminal solid solution can be determined as a function of solidification velocity. Figure 1.13 schematically shows this dependence. In the figure, the ratio of the actual composition of the crystal to its equilibrium value $K = X_s/X_{eq}$ is called the partitioning coefficient. It has the value $K = 1$ in equilibrium, while $K > 1$ for a non-equilibrium composition. $\ln K$ is plotted as a function of solidification velocity.

The Cu–Ti phase diagram is shown in Fig. 1.14. Here, the T_0 lines are

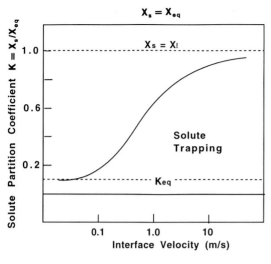

Fig. 1.13. Ratio of the composition of a growing crystal to that of the parent liquid from which it grows as a function of solidification velocity. This ratio is called the solute partitioning coefficient, K. Its value approaches unity at high cooling rates (high solidification velocities). (See [1.18] for details)

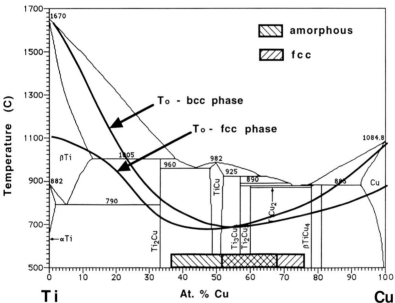

Fig. 1.14. Phase diagram of Cu–Ti showing T_0 lines of various phases as well as the regions in which metastable amorphous and fcc phases are formed

shown for the terminal fcc Cu-phase as well as for the hcp α-Ti and bcc β-Ti terminal phases. In addition, the T_0 line is shown for the intermetallic compound Ti$_2$Cu. The T_0 lines were determined by *Massalski* [1.24]. Rapid solidification in this system produces a wide variety of metastable phases [1.24]. At very high cooling rates ($\approx 10^6$ K/s), a glassy phase is produced over the central portion of the phase diagram, as indicated in the figure. Under these conditions, no crystal nucleation occurs. At lower cooling rates or at compositions lying outside the glass forming region, a metastable fcc solid solution is formed during solidification. This metastable fcc phase is formed at compositions, which, in equilibrium, consist of an intermetallic compound or a mixture of two intermetallic compounds. Here, the extension of the solubility of the fcc Cu phase displaces the equilibrium intermetallics. The metastable fcc phase is now found at compositions where it does not occur at all in the equilibrium system, even as part of a two-phase mixture. To understand this behavior, one must assume that nucleation of the fcc phase is kinetically favored over that of the intermetallics. In particular, during rapid quenching and the accompanying undercooling, the melt must undercool below the T_0 line of the fcc phase. Furthermore, the nucleation barrier for the fcc phase at such undercoolings must be lower than that for the intermetallic compounds. Since the intermetallics are necessarily at a larger undercooling (the equilibrium liquidus curves of the intermetallics are at higher temperature than the liquidus of the fcc phase), one expects the driving force for crystallization of the intermetallic to be larger than that of the fcc phase. According to Sect. 1.2, this suggests that the interfacial energy of the fcc–liquid interface is lower than that of the intermetallic–liquid interface. Only then would the nucleation barrier be lower.

1.4.3 Metastable Crystalline Phases Not Present in Equilibrium – Examples

In Sect. 1.2.3, the formation of a metastable θ-phase was illustrated in Figs. 1.5 and 1.6. According to the discussion there, this phase is completely absent in the equilibrium diagram Fig. 1.5 but is favored during rapid solidification. It could be favored for one of two reasons. The composition shift required to nucleate the competing crystalline phases may be kinetically sluggish as outlined in Sect. 1.2, or the interfacial tension of the θ-phase–liquid interface might be significantly lower than that of the β-phase– liquid interface. Either of these factors could result in a lower nucleation barrier and the preferred nucleation of the θ-phase during rapid solidification and accompanying deep undercooling. The metastable θ-phase completely replaces the equilibrium phases. This type of new phase is found surprisingly often in rapid solidification experiments. Table 1.1 gives a number of selected examples which are illustrative and interesting. In equilibrium, the first two examples, Ag–Ge and Ge–Sb, are simple eutectic systems with no intermetallics. They are of the type illustrated in Figs. 1.5 and 6. During rapid solidification of Ag–Ge (containing only the fcc Ag and diamond cubic Ge phases in equilibrium), one finds a new metastable hcp (A3-type) phase at 15–26 at.% Ge [1.25]. In the Ge–Sb eutectic system, rapid quenching results in a new π-phase [1.26]. The π-phase is an unusual structure. The unit cell contains a single atom and is primitive cubic (although the cube is slightly distorted in Ge–Sb). The Ge and Sb atoms are randomly situated on the simple cubic lattice. Among the elements, this simple but uncommon structure is only found in Polonium (α-Polonium structure). It is frequently found as a metastable phase in rapidly quenched alloys [1.26, 27].

The simple "Cs–Cl-type" structure (B2-type) is quite common in equilibrium phase diagrams of rare-earth metals with late transition metals. Nonetheless, it is notably missing in a number of phase diagrams where other competing phases form instead. For example, the B2 phase is absent in most Au–Re phase diagrams (RE = rare-earth metal). Rapid solidification has been found to produce a metastable B2 phase in many of these systems. Examples are shown in Table 1 and include Au–Y, Au–Pr, Au– Nd, etc. Presumably, the occurence of the B2 phase in rapid quenching experiments is related to the relative ease with which this phase nucleates by comparison with the competing equilibrium phases. Since the B2 phase is absent in the equilibrium phase diagram of these systems, it can be said that rapid solidification has produced a strictly new phase.

There are many other examples of non-equilibrium crystalline phases which are produced using the rapid solidification technique. The reader is referred to a review article by Jones [1.28] for a comprehensive summary of these examples.

Table 1.1. Some examples of new metastable phases (not present in the equilibrium phase diagrams) produced by rapid solidification

Alloy System	Equilibrium phases	Metastable phase	Ref.
Ag–Ge	Simple eutectic	hcp phase, A3-type 15–26 at.% Ge	1.25
Au–Ge	Simple eutectic	hcp phase, A3-type 25 at.% Ge	1.25
Au–Te	Two eutectics with $AuTe_2$	Simple cubic π-phase (1 atom/unit cell) 55–90 at.% Te	1.25
Ge–Sb	Simple eutectic	Simple cubic π-phase (distorted)	1.26
Pd–Sb	Many intermetallics	Simple cubic π-phase (16 at.% Pd)	1.26
Au–Y, Au–Pr, Au–Nd, Au–Sm, Au–Gd, etc.	Many intermetallics	Cs–Cl-type, (B2) 50 at.% Au	1.29
Cu–Th, Fe–Th	Many intermetallics	CrB type (B33) 50 at.% Th	1.30
Cu–Ti	Cu_3Ti equilibrium phase	DO_2-type	1.30
Au–Sn	Four intermetallics	$D8_2$-type (γ-brass)	1.26

To summarize, rapid solidification tends to restrict the kinetic processes of crystal-phase nucleation and growth from the melt so as to prevent the attainment of equilibrium. Among other things, this leads to solidification at relatively deeper undercoolings compared with conventional processing. Under such circumstances, metastable phases having melting points which lie below the equilibrium liquidus of the system become favored by rapid solidification when the nucleation barrier to the metastable phase is lower than that of the competing equilibrium phase.

Thus, the products of rapid solidification are various metastable crystalline alloys and compounds. If no such metastable or stable phase can be nucleated during the brief time of quenching, a glass is obtained. In this book various properties of these products of rapid solidification, metallic glasses in particular, are introduced, and their applications are discussed.

References

1.1 *Rapidly Solidified Alloys: Processes, Structures, Properties, Applications*, ed. by H.H. Liebermann (Dekker, New York 1993)
1.2 *Science and Technology of Rapid Solidification and Processing*, ed. by M.A. Otooni (Kluwar, Dordrecht 1995)
1.3 E.g., D.A. Porter, K.E. Easterling: *Phase Transformations in Metals and Alloys* (Chapman & Hall, London 1981)
1.4 T.B. Massalski, *Proc. 4th Intl. Conf. Rapidly Quenched Metals*, ed. by T. Masumoto, K. Suzuki (Jpn. Inst. Metals, Sendai 1982) p. 203
1.5 K. Klement, R.H. Willens, P. Duwez: Nature **187**, 869 (1960)
1.6 D. Turnbull: Contemp. Phys. **10**, 471 (1969)
1.7 T. Egami, Y. Waseda: J. Non-Cryst. Solids **64**, 113 (1984)
1.8 E.g., C. Suryanarayana: *A bibliography 1973–1979 for Rapidly Quenched Metals* (Plenum, New York 1980)
1.9 Y. Waseda, S. Ueno, M. Hagiwara, K.T. Aust: Progr. Mater. Sc. **34**, 149 (1990)
1.10 *Glassy Metals I, Ionic Structure, Electronic Transport, and Crystallization*, in Topics Appl. Phys., Vol. 46, eds. H. Beck and H.-J. Güntherodt, Topic (Springer Verlag, Berlin, 1981); *Glassy Metals II, Atomic Structure and Dynamics, Electronic Structure, Magnetic Properties*, in Topics Appl. Phys., Vol. 53, eds. H. Beck and H.-J. Güntherodt, Topic (Springer Verlag, Berlin, 1983); *Glassy Metals III, Amorphization Techniques, Catalysis, Electronic and Ionic Structure*, in Topics Appl. Phys., Vol. 72, eds. H. Beck and H.-J. Güntherodt, Topic (Springer Verlag, Berlin, 1994).
1.11 A. Peker, W.L. Johnson: Appl. Phys. Lett. **63**, 2342 (1993)
1.12 W.L. Johnson: Mater. Sci. Technol. **9**, 94 (1994)
1.13 A. Inoue: Mater. Trans. JIM **36**, 866 (1995)
1.14 A.L. Greer: In *Rapidly Solidified Alloys*, ed. by H.H. Liebermann (Dekker, New York 1993) p. 269
1.15 T. Egami: In *Amorphous Metallic Alloys*, ed. F.E. Luborsky (Butterworths, London 1983) p. 100
1.16 B.E. Warren: *X-ray Diffraction* (Dover, New York, 1990)
1.17 J.D. Bernal: Nature **188**, 410 (1960)
1.18 G.D. Scott: Nature **188**, 408 (1960)
1.19 F.C. Frank: Proc. Roy. Soc., London A **215**, 43 (1952)

1.20 D. Shechtman, I. Blech, D. Gratias, J.W. Cahn: Phys. Rev. Lett. **53**, 1951 (1984)
1.21 E.g., *Quasicrystals, The State of the Art*, ed. by D.P. DiVincenzo, P.J. Steinhart (World Scientific, Singapore 1991)
1.22 P. Duwez: Fizika, Suppl. **2**, 1 (1970)
1.23 M.J. Aziz, T. Kaplan: Acta Metall. **36**, 2335 (1988)
1.24 T.B. Massalski: In *Rapidly Quenched Metals*, ed. by S. Steeb, H. Warlimont (Elsevier, Amsterdam 1985) p. 171
1.25 P. Duwez: In *Intermetallic Compounds*, ed. by J.H. Westbrook (Wiley, New York 1967) Chap. 18
1.26 B.C. Giessen: Z. Metallk. **59**, 805 (1968)
1.27 W.L. Johnson, S. Poon: J. Appl. Phys **45**, 3683 (1974)
1.28 H. Jones: *Rapid Solidification of Metals and Alloys* (Chameleon, London 1982), H. Jones: Rep. Prog. Phys. **36**, 1425 (1973)
    H. Jones: Mater. Sci. Eng. A **179/180**, 1 (1994)
1.29 C.C. Cao, H.L. Lue, P. Duwez: J. Appl. Phys. **35**, 247 (1964)
1.30 B.C. Giessen: In *Rapidly Quenched Metals II*, ed. by B.C. Giessen, N.J. Grant (MIT Press, Boston 1976) ps. 119, 149

2 Synthesis and Processing

N.J. Grant, H. Jones, and E.J. Lavernia

Rapid Solidification (RS) involves propagation of a solidification front at high velocity. This is most readily achieved by suitable treatment of a volume of melt. Suitable treatments include: (*i*) dividing it up into a multitude of small droplets (atomisation, emulsification or spray-forming) so that most of them can undercool deeply prior to solidification; (*ii*) stabilising a meltstream of small cross section in contact with an effective heat sink (melt-spinning or thin-section continuous casting); (*iii*) rapid melting of a thin layer of material in good contact with an extensive heat sink, which may be the same or related material (electron or laser beam surface pulse or traverse melting). In each case rapid solidification results from rapid extraction of the heat of transformation either directly by the external heat sink and/or internally by the undercooled melt (in which case the system rapidly reheats, i.e., recalesces during solidification). The large undercoolings developed amount to large departures from equilibrium leading to formation of extended solid solutions and new non-equilibrium phases (crystalline, quasicrystalline or glassy) while the short freezing times give rise to size-refined and compositionally rather uniform microstructures as well as relatively high rates of throughput of material. The products of RS range from powder or flake particulate, through thin discontinuous or continuous ribbon or filament to thick spray deposits containing some trapped porosity. These products can sometimes be applied directly as in the cases of finely divided light metal particulate used as the basis for space shuttle and satellite launch rocket fuel and signalling flares, and planar-flow-cast strip used in certain magnetic applications or for braze assembly of engine components. For most applications, however, they must be suitably incorporated or consolidated into full size, fully dense sections or components. This may involve processes such as polymer bonding or liquid metal infiltration but most commonly involves powder metallurgy techniques such as die or isostatic pressing and/or hot working.

The present chapter is devoted to setting out the principles underlying RS technology. Earlier reviews of this process technology have been published by *Duwez* [2.1], *Jones* [2.2–2.4], *Savage* and *Froes* [2.5], *Anantharaman* and *Suryanarayana* [2.6] and by *Suryanarayana* [2.7].

2.1 Heat Transfer and Solidification Kinetics

A fundamental feature of all rapid solidification processing is that heat evolved during solidification must be transferred with sufficient rapidity to a heat sink. For an undercooled melt, the first effect of crystalline solidification is to raise the temperature of the system until solidification is complete or until the solidus temperature is reached, whichever occurs first. This process of recalescence is a typical feature of cooling curves of solidifying melts. During recalescence, the rate of evolution of latent heat in the system exceeds the capability of the external heat transfer process to remove it to the surroundings. The rate of advance of the solidification front during this stage can be very rapid indeed (velocities as high as 100 m/s have been measured [2.8]) but residual freezing following its completion occurs at a much slower rate governed by the operative rate of heat transfer to the surroundings. This two-stage process can account for the two-zone structure often observed in samples of rapidly solidified materials. An example is shown in Fig 2.1 in which the featureless single-phase region is deemed to have formed during recalescence while the microduplex region formed during the subsequent stage of heat extraction to the surroundings [2.9].

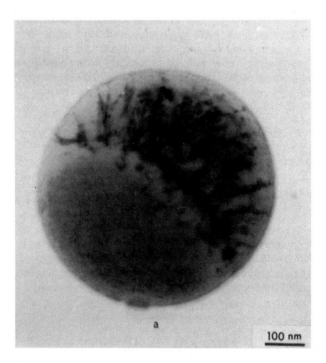

Fig. 2.1. Showing transition from featureless single phase to duplex microstructure in an electrohydrodynamically atomized Al-6 wt.% Si alloy rapidly solidified droplet. Transmission electron micrograph from [2.9]

The proportion of such undercooled material can be used to estimate the level of undercooling at which solidification was triggered. One hundred percent of such material indicates a nucleation undercooling of at least L/c, where L is the latent heat solidification and c the specific heat of the melt, at the onset of solidification while a fraction f of such material indicates a nucleation undercooling of fL/c. High-resolution electron microscopy of the 'featureless' material often demonstrates that it is at least partially cellular with some intercellar second phase which may be a glass. In such cases (and also for the microduplex region), the spacing and solute concentration of the cells can be related to predictions of cellular growth at high velocity, with reasonable agreement being obtained [2.10].

The situation in which the product of rapid solidification is entirely or primarily a glass, is related in that continuous cooling of the whole melt down to T_g, the glass formation temperature, must be achieved before the competitive process of crystal formation described above can start or at least advance significantly. Undercooling is an important consideration, therefore, whether or not recalescence occurs and cooling rate, in particular, thereby plays a pivotal role in all these situations.

Visualize (Fig. 2.2) a molten sphere of radius r travelling in a cool gaseous medium (typical of droplet RS processes), a molten cylinder of radius r injected into a bath of liquid coolant (as occurs in the production of rapidly solidified wire) or a parallel-sided slab of melt of thickness z in at least partial contact on one side with a chill substrate (as in splat quenching, Sect. 2.2 or chill-block melt spinning RS processes, Sect. 2.3). Assume that heat is lost to the gas, liquid coolant or chill-block at temperature T_A at a rate governed by an interfacial heat transfer coefficient h. Let the initial uniform temperature of the melt be T_B and assume that it freezes on reaching temperature T_F. Assume for now that h is sufficiently low to ensure that the temperature remains essentially uniform

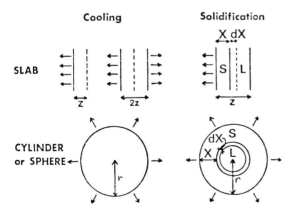

Fig. 2.2. Cooling and solidification of a sphere or cylinder, or of a slab. Arrows show direction of external heat extraction. S = solid, L = liquid. [2.3]

throughout the sphere or slab during cooling and freezing, i.e., that 'Newtonian' conditions apply. To evaluate the cooling rate $\dot{T}$ at any temperature T, equate the specific heat released during a small decrease in temperature dT to the heat removed in a corresponding time increment dt, i.e.,

$$c\rho V_o(-dT) = hA_o(T - T_A)\,dt,$$

where V_0 is the volume and A_0 the surface area losing heat and ρ is the density. Thus,

$$\dot{T} = -\frac{dT}{dt} = \frac{h(T-T_A)}{c\rho}\frac{A_o}{V_o}, \qquad (2.1)$$

where A_0/V_0 is $3/r$, $2/r$ and $1/z$, respectively, for a sphere, cylinder or slab. Correspondingly, to evaluate the average solidification front velocity $\dot{X}$ that is directly sustainable by the heat transfer coefficient h, equate the latent heat released at temperature T_F during an incremental advance dX in solidification front position X, again to the heat removed in a corresponding time increment dt, i.e.,

$$L\rho A_F\,dX = hA_o(T_F - T_A)\,dt,$$

where A_F is the instantaneous area of the solidification front at position X so that

$$\dot{X} = \frac{dX}{dt} = \frac{h(T_F - T_A)}{L\rho}\frac{A_o}{A_F}, \qquad (2.2)$$

where A_0/A_F is $r^2/(r-X)^2$ for the sphere, $r/(r-X)$ for the cylinder and unity for the slab. Thus, both cooling rate $\dot{T}$ and front velocity $\dot{X}$ are proportional to h and, in addition, $\dot{T}$ is inversely proportional to radius r or thickness z. Both $\dot{T}$ and $\dot{X}$ are independent of position X except for the sphere or cylinder for which $\dot{X}$ increases steadily from $h(T_F - T_A)/L\rho$ initially to infinity at completion of solidification.

Equations (2.1 and 2.2) for Newtonian conditions apply when the Nusselt number Nu $= hr/k$ or $hz/k \ll 1$, where k is the thermal conductivity of the solidifying material. At the other extreme Nu $\gg 1$, 'ideal' cooling conditions prevail in which temperatures, cooling rates and front velocities then depend on position X. Analysis for a slab gives, corresponding to (2.1 and 2.2)

$$\dot{T} = \alpha/X^2 \qquad (2.3)$$

and

$$\dot{X} = \beta/X, \qquad (2.4)$$

where α, β are functions of the relevant temperature intervals and materials properties but are now independent of h. Typical values of α and β for a metallic melt cooling and freezing against a metal chill are $\approx 10^4\,\text{mm}^2\,\text{K/s}$ and $10^2\,\text{mm}^2/\text{s}$, respectively [2.11], Equations (2.3 and 4) then indicating the upper limits of $\dot{T}$

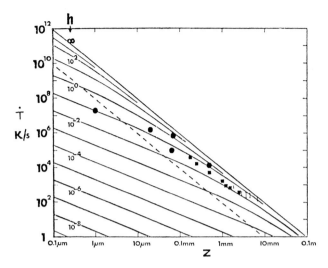

Fig. 2.3. Predicted average cooling rate $\dot{T}$ in the solidification range as a function of thickness z and heat transfer coefficient h (W/mm² K) for aluminium freezing against a copper chill, compared with experimental measurements, represented as points [2.2]

and $\dot{X}$ attainable at the more representative *unchilled* surface $X = z$ of the slab. How nearly these limits are approached depends on the value of Nu $= hz/k$ so that a higher value of h is required for a thinner section. This is illustrated in Fig. 2.3 for aluminium cooling and freezing on a copper chill, showing that at $z = 10$ mm, an h of 10^{-2} W/mm² K is sufficient to give $\dot{T}$ within a factor of 10 of the ideal 10^4 K/s, while, at $z = 1$ μm, h must be 10^4 times higher to give $\dot{T}$ within this same factor of the ideal 10^{10} K/s. For the given conditions of thermal contact with or heat transfer to the surroundings, the inevitable result is that there is always a threshold value of z below which $\dot{T}$ and $\dot{X}$ are controlled by the magnitude of h. The magnitude of h depends on the mechanisms of heat transfer.

Radiative heat transfer gives

$$h = \varepsilon \sigma T^3 \tag{2.5}$$

provided that $(T - T_A)$ is not small, where ε is the emissivity and σ is Stefan's constant $(5.8 \times 10^{-14}$ W/mm² K⁴). For $T = 1000$ K and $\varepsilon = 1$, thus $h \approx 10^{-4}$ W/mm² K for radiation cooling so that cooling is always Newtonian for z below 10 mm.

Convective cooling, e.g., of a sphere of radius r moving with relative velocity u in a conducting gas gives

$$\text{Nu} = 2 + 0.6 \, \text{Re}^{1/2} \, \text{Pr}^{1/3}$$

where Nu $= 2rh/k$, the Reynolds number Re $= 2ru\rho/\eta$ and the Prandtl number Pr $= c\eta/k$, and k, ρ, c and viscosity η are parameters for the gas. Thus,

$$h = \frac{k}{r} + 0.6 \left(\frac{k^4 \rho^3 c^2}{\eta}\right)^{1/6} \left(\frac{u}{2r}\right)^{1/2}. \quad (2.6)$$

For a sphere of diameter 100 μm travelling at 500 m/s relative to helium gas, (2.6) gives $h \approx 10^{-2}$ W/mm² K so that cooling will be Newtonian at this or smaller u or for smaller r at this u.

Conduction through a thin insulating film, e.g., of a gas or oxide between melt and chill surface, gives

$$h = k/\delta, \quad (2.7)$$

where k is the film conductivity and δ is its thickness. For a film of air of thickness 0.5 μm, this gives $h \approx 0.1$ W/mm² K representative of experimental measurements for contact with chill surfaces typical of most methods of rapid solidification processing (Fig. 2.3), and gives Newtonian cooling only when $z < 10$ μm.

The practical conditions for rapid solidification at a high cooling rate are therefore very clear. A volume of melt with at least one dimension that is small must be rapidly formed in good contact with an effective heat sink. The required size of this small dimension depends on the cooling rate desired. For example, it is impossible to achieve $> 10^4$ K/s in a thickness > 1 mm or to reach $> 10^8$ K/s in a thickness > 10 μm and the actual cooling rate will decrease linearly with decreasing h if h falls below $\approx 10^{-3}$ and 10^{-1} W/mm² K, respectively, in the two cases.

Two examples of the importance of the cooling rate in determining the outcome of rapid solidification processing are given in Figs. 2.4 and 2.5. Figure

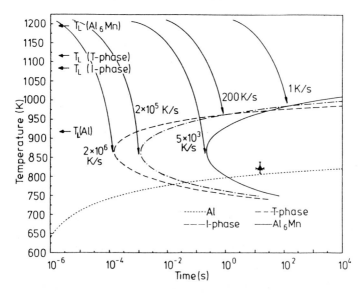

Fig. 2.4. Predicted temperature-time-transformation curves for competing phases in solidification of undercooled Al-12 at.% Mn alloy indicating threshold cooling rates in K/s required to form each phase [2.12]. Liquidus temperatures T_L of the competing phases are indicated

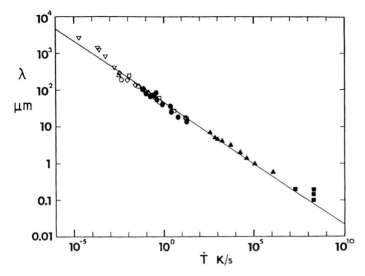

Fig. 2.5. Dendrite cell size λ of α-Al phase as a function of cooling rate $\dot{T}$ for Al-4 to 5 wt.% Cu and Al-7 to 11 wt% Si alloys. Best fit line shown represents $\lambda \dot{T}^{1/3} = 50\,\mu\mathrm{m}\,(\mathrm{K/s})^{1/3}$ [3.2]

2.4, based on specific assumptions regarding the nucleation and growth kinetics, shows that cooling rates exceeding 2×10^6, 2×10^5 and 5×10^3 K/s are predicted to be required to suppress formation of I, T and $\mathrm{Al}_6\mathrm{Mn}$ primary phases as the outcome of solidification of Al-12 at.% Mn alloy melts, values in good accord with experimental observations [2.12]. Figure 2.5 represents an example of the empirical relationship

$$\lambda = a\dot{T}^{-n} \qquad (2.8)$$

between dendrite cell size and cooling rate $\dot{T}$ during solidification established for a number of systems. Typically $0.25 < n < 0.5$ and $20 < a < 160\,\mu\mathrm{m}\,(\mathrm{K/s})^n$, not too different from predictions of volume-controlled coarsening theory which gives $n = 0.33$ and, for the Al–Cu and Al–Si alloys represented in Fig. 2.5, $a \approx 50\,\mu\mathrm{m}\,(\mathrm{K/s})^{1/3}$ [2.13].

2.2 Droplet Methods[1]

Typically, these make use of the fact that a pendant drop or free falling melt stream tends to break up into droplets as a result of surface tension. This

---

[1] For recent monographs featuring this subject area see A. Lawley: *Atomization: The Production of Metal Powders*, MPIF, Princeton, NJ 1992. A.J. Yule and J.J. Dunkley: *Atomization of Melts for Powder Production and Spray Deposition*, Clarendon Press, Oxford 1994. E.J. Lavernia and Y. Wu: *Spray Atomization and Deposition*, Wiley, New York 1996.

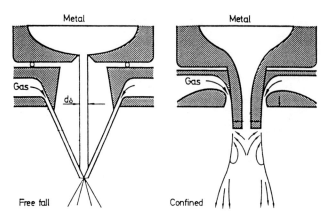

Fig. 2.6. Principle of spray droplet formation by impingement of high velocity gas jets on a free-falling or emerging meltstream. [2.14]

process can be intensified, for example, by impingement of high-velocity jets of a second fluid, by centrifugal force at the lip of a rotating cup or disc or by an applied electric field. The standard method uses high-velocity gas or water jets (Fig. 2.6) [2.14]. These form, propel and cool the droplets that may freeze completely in flight or may form individual splats or build up a thick deposit by impact on a suitable substrate. Powder or shot particulate generated by such techniques displays a range of particle sizes and shapes even for a single set of operating conditions with increasing atomizing gas pressure, for example, increasing the yield of finer particle sizes and the more effective quench of water atomization leading to more irregular particle shapes. Although particles of more or less identical size and shape from the same powder sample can show quite different microstructures, smaller particles tend to cool more rapidly and/or undercool more prior to solidification so tend to solidify more rapidly (i.e., at higher front velocity). Splats formed from droplets of a given size tend to solidify even more rapidly because of more effective heat extraction from the larger surface area they offer, especially when at least one of their surfaces is in good contact with an efficient heat sink, such as a water-cooled rotating copper drum. A spray deposit can maintain the same microstructure as the equivalent splats provided that their solidification time is sufficiently less than the interval between deposition of successive splats at a given location on the substrate [2.15]. Atomizers range in size from laboratory units with capacities of less than 1 kg per run to commercial size facilities with capacities as large as 50 000 tonnes per year [2.16]. Powders destined for high-performance applications tend to be atomized with inert gases or in vacuum to minimize formation of oxides or other potentially damaging inclusions.

Except for very well-defined situations, such as successive separation of single droplets from a pendant source of known dimensions, the fundamental mechanisms governing droplet formation in atomization processes remain to be established. It is widely accepted, however, that Rayleigh instabilities play a

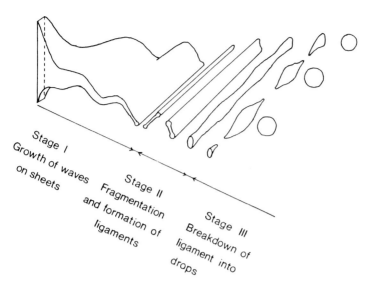

Fig. 2.7. Classical model of breakup of a moving liquid sheet first into ligaments and then into droplets [2.17]

crucial role. This is illustrated in Fig. 2.7 in which these instabilities lead to the breakup of liquid sheets first into ligaments which then pinch off into lozenge-shaped short filaments which in turn spherodize into droplets [2.17]. Any particularly large droplets generated by this process are likely to be unstable in the velocity field and undergo secondary breakup by first becoming flat, then bowl-shaped, finally bursting into a number of smaller droplets. Complete or partial coalescence of droplets can also result from collisions between droplets at different stages of solidification. A correlation originated by *Lubanska* [2.18] gives the mass mean droplet diameter d_{50} produced by high-velocity gas jet atomization as

$$d_{50} = M\left(\frac{\eta_m \gamma_m d_o}{\eta_g \rho_m v^2}\right)\left(1 + \frac{J_m}{J_g}\right)^{1/2}, \tag{2.9}$$

where $40 < M < 400$, η, γ, ρ are viscosity, surface tension and density, d_o is the diameter of the melt-stream delivery nozzle, v is the relative velocity of the atomizing gas, J the mass flow rate, and suffixes m and g refer to the melt and gas, respectively. In situations where there is sufficient time for the droplets to complete freezing before they strike the wall of the atomizing chamber, their shape is governed by competition between spherodization and solidification. The spherodization time is given [2.19] by $3\pi^2 \eta (r_1^4 - r_0^4)/4V_0\gamma$, where η and γ are the viscosity and the surface tension of the melt, V_0 is the volume of the particle, and r_0 and r_1 are its initial and final radii. The spherodization time of an aluminium droplet of diameter 80 μm is thereby predicted to be no more than 10^{-5} s. This compares with a solidification time $L\rho r/3h(T_F - T_A)$ by integration of (2.2) for a sphere. For $h = 10^{-1}$ W/mm² K, this gives $\approx 10^{-3}$ s, giving

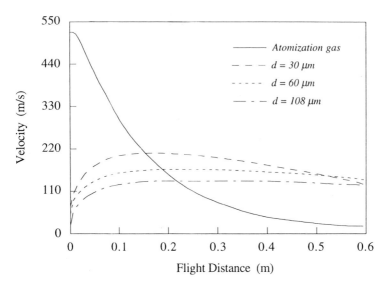

Fig. 2.8. Predicted velocity profiles along the axis of atomization for N_2 atomized Ni_3Al droplets of three different sizes d, and for the atomizing gas [2.20]

ample time for spherodization prior to solidification even for this relatively large droplet size. Droplets are accelerated initially by the atomizing gas up to a maximum velocity of ≈ 100 m/s over a distance of ≈ 100 mm for typical conditions (Fig. 2.8) [2.20]. Because the effective convective heat transfer coefficient h depends on the instantaneous *difference* in velocity between the atomizing gas and the droplet, h reaches a minimum when the droplet reaches its maximum velocity (Fig. 2.9) [2.21]. A typical computed solidification history for such an aluminium droplet is shown in Fig. 2.10 [2.22].

Spray deposition involves the impact and coalescence on a solid surface of a distribution of droplets. These typically have different sizes and velocities, and are at different stages of cooling and solidification, from fully liquid, through partially solid to fully solid. The structure of the resulting deposit can thus vary from what would result from continuous solidification of a thick liquid layer to the aggregates of a series of independent solidification events associated with the spreading and interaction of individual droplets on the deposition surface. The former situation (A) is favoured by high rates of deposition onto a deposition surface which is near to its solidus temperature, conditions not too dissimilar from those of investment casting. The latter situation (B) is favoured by low rates of deposition of highly undercooled (but not solidified) high-velocity small droplets onto a chilled deposition surface. Situation A would tend to favour near-equilibrium slow freezing to produce coarser microstructures relatively free of trapped porosity. Situation B, in contrast, favours rapid freezing to produce highly non-equilibrium layered microstructures with an increased possibility of entrapment of gas porosity between the layers. Typical conditions for

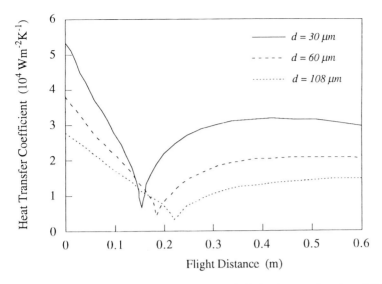

Fig. 2.9. Calculated heat transfer coefficient of Ni_3Al droplets as a function of flight distance and droplet size [2.21]

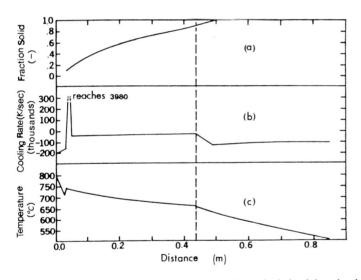

Fig. 2.10. Computed solidification history of an 80 μm-sized aluminium droplet [2.22]

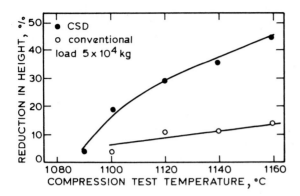

Fig. 2.11. Improvement in hot-working response of Controlled Spray Deposited (CSD) M2 tool steel compared with conventional ingot-derived M2 [2.23]

the production of large preforms by spray deposition tend to be closer to situation A than situation B. A critical feature, however, is the presence in the dispersion of droplets impacting on the deposition surface of some 40 to 90% of already solidified material which results in the uniform macrosegregation free non-dendritic fine equiaxed microstructures that are so characteristic of such preforms. Such microstructures show excellent hot deformation characteristics (Fig. 2.11) [2.23] and thus provide an ideal feedstock material for shaping of materials such as tool steels or metal matrix composites at low forging pressures in their partially remelted semi-solid condition [2.24].

The complexity and variety of individual events in the initial formation of the spray, in its travel to the deposition surface and on the deposition surface

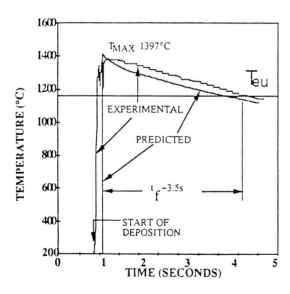

Fig. 2.12. Predicted temperature versus time profile compared with experimental results for spray-deposited steel, 2.5 mm into the growing deposit [2.25]. T_{Eu} = eutectic temperature; t_f = solidification time

itself all provide a challenge to the experimentalist as well as a field day for the numerical process modeller. The evolution of droplet size distribution and velocity in the travelling spray can be monitored intrusively or non-intrusively by collecting droplet samples, by short duration exposure photography and laser Doppler anemometry. Thermal history of the growing deposit can be followed by infrared thermography of its surface and via embedded thermocouples. Metallographic observations on finished deposits such as the diameter of the largest entrained presolidified droplet, magnitudes of dendrite cell sizes or eutectic spacings and occurrence of non-equilibrium phases, all provide internal checks on the validity of predictive numerical process models. In spite of the inevitable simplifications built into such models, the level of agreement between predictions and experimental fact is encouraging. An example of such a comparison is shown in Fig. 2.12 [2.25].

2.3 Spinning Methods[2]

Spinning methods derive from the simplest system, in which a single melt stream emerging from an orifice is stabilized by surface film formation or solidification before it can break up into droplets (so-called *melt extrusion* or *melt spinning*), to the most sophisticated in which one or more melt streams are used to make wide or composite ribbons by impingement on single or twin chill roll surfaces. In standard *free-jet chill-block melt spinning* (Fig. 2.13) [2.26] the meltstream forms a single ribbon typically 10 to 100 µm thick and a few millimetres in

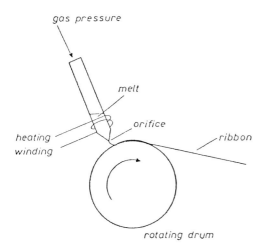

Fig. 2.13. Schematic of free-jet chill-block melt-spinning [2.26]

---

[2] Two recent Conference Proceedings feature these methods: *Melt-spinning and Strip Casting*, ed. by E.F. Matthys, TMS, Warrendale, 1992. *Melt-spinning, Strip Casting and Slab Casting*, ed. by E.F. Matthys and W.G. Truckner, TMS, Warrendale, PA 1996.

width by impingement onto a single chill roll rotating at a surface speed of some tens of metres per second. In *planar flow casting* the distance from the nozzle orifice to the chill surface is reduced to less than ≈0.5 mm to eliminate formation of the melt pool and associated instabilities. Thin strips up to 500 mm in width can be cast in tonnage quantities by this method. *Melt-extraction*, in which the rotating chill roll forms the product by direct contact with the surface of a crucible of melt or the melted extremity of the solid electrode, and its derivative, *melt overflow*, do not involve generating a melt stream. Melt-extraction is well established as a cost-effective technique for producing tonnage quantities of staple fibres for reinforcement of concrete or refractories [2.27]. Continuous rapidly solidified filament with round section can be produced by the *rotating water bath* process in which a melt stream emerging from a cylindrical orifice is solidified directly on entry into a rotating annulus of water contained within the lip of a rotating water bath [2.28]. This process can also be used to generate powder particulate when the conditions result in break-up of the melt stream on entry into the water bath [2.29]. Free-jet chill-block melt spinning and melt-extraction be used to generate a flake product directly if continuity is interrupted by a series of regularly spaced notches on the chill-block. Such (aluminium) flake can be loaded into a polymer matrix to combine the conducting properties of a metal with the processing economies and flexibilities offered by plastic [2.30]. Melt-spun ribbon or planar flow cast sheet can be pulverized, for example, by means of a blade cutter mill [2.31, 32] into platelets ≈0.3 mm across. These provide a more suitable feedstock for consolidation by powder metallurgy than continuous ribbon, for example, which exhibits a much lower packing density.

As for other rapid solidification methods, minimum cross-sectional dimension and the operative heat transfer coefficient h at surfaces are important in determining the structure and properties of products of spinning. In certain configurations such as melt extrusion and the rotating water bath process, section dimensions of a continuous product are limited primarily by expulsion orifice dimensions, while the operative h depends on jet velocity relative to that of any quenching medium employed. When impingement onto a solid chill surface is involved, a higher relative velocity of the chill surface can result in substantial further thinning of the smallest section dimension as well as increased operative values of h. The most widely-used of these processes, Free-Jet Chill-Block Melt-Spinning (FJCBMS) is the most complex in these respects in that resultant ribbon thickness, width and contact length with the chill surface are all dependent on process variables. Early work by *Kavesh* [2.33] for metallic glass ribbons made by FJCBMS showed a thickness z and width w typically proportional to $Q^{0.2}/v^{0.8}$ and $Q^{0.8}/v^{0.2}$, respectively, where Q is the volumetric melt flow rate and v is the chill-block surface speed. Modelling based on limitation of z by momentum boundary layer formation at the chill surface boundary of the melt puddle can give good agreement with experimental data, by making reasonable assumptions concerning magnitudes of the temperature dependence of melt viscosity and of h at the boundary (Fig. 2.14) [2.34]. For the equivalent

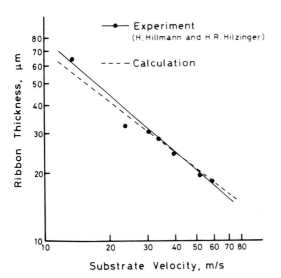

Fig. 2.14. Ribbon thickness z versus chill surface velocity v for free-jet chill-block melt-spinning of Fe-40at.% Ni-14at.%P-6at.%B metallic glass, showing $z \propto v^{-0.8}$ [2.34]

situation in planar flow casting, where the ribbon width is constrained to match the orifice slot width, the ribbon thickness is determined by continuity of melt flow and the Bernouilli equation to give

$$z = b(w_o/v)(2\Delta P/\rho)^{1/2}, \tag{2.10}$$

where w_0 is the orifice width, v the chill surface speed, ΔP the pressure difference driving expulsion of the melt via the orifice, ρ the density of the melt and b is a numerical constant, ranging in practice between 0.6 and 0.7 [2.35]. An example of agreement with (2.10) is shown in Figs. 2.15a,b [2.36]. The chill surface acts as an effective heat sink for the cooling and solidification processes only as long as contact is maintained with it. While this contact length may be several centimetres in chill-block melt spinning or planar flow casting it may be only a few millimetre in twin roll casting of thin sheet in which contact with at least one of the chill rolls must cease at the point of minimum separation of the roll surfaces. Separation from the chill surface typically results in a substantial decrease in operative h so that the cooling rate reduces sharply from that point onwards. The fact that this drop in h does not appear to be deleterious probably reflects the marked reduction in diffusivity with decrease in temperature that occurs even for glasses, ensuring that the beneficial effects of rapid solidification are not reversed by a somewhat slower cool following completion of solidification.

Many products of this group of processes are used in applications where surface quality of the as-cast product is important, in addition to the intrinsic properties of the material itself. Entrainment of gas at the melt–pool/chill–roll interface reduces the local h and can lead to formation of elongated grooves on the chill surface of the ribbon. Local instabilities in fluid flow can also give rise to herring-bone patterns on the chill side. Similarly, unevenness on the free

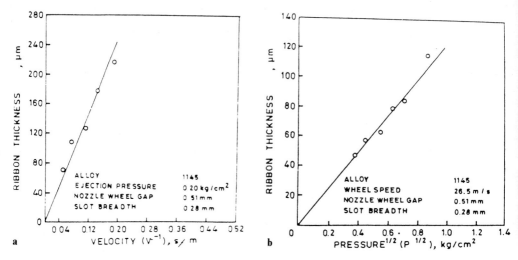

Fig. 2.15. Ribbon thickness z versus (a) inverse of chill surface velocity v and (b) square root of injection pressure P for planar flow casting of 1145 aluminium alloy [2.36]

surface of the ribbon can result, for example, from interdendritic feeding of solidification shrinkage, as for larger atomised droplets, under conditions where crystalline solidification results. Consistent achievement of adequate surface quality as well as good dimensional control across the width and along the length of a continuous thin strip produced by twin-roll casting are critical requirements for widespread introduction of this process for industrial-scale production of feedstock for finishing by cold rolling, in competition with conventional slab casting followed by hot rolling.

2.4 Surface Melting Technologies

Surface melting technologies derive from spot or continuous welding techniques and differ from them only in that the depth melted is limited to ensure that the ensuing solidification will be sufficiently rapid. In its simplest form a single pulse or continuous traversing heat source is used to rapidly melt the surface of a block of material, the unmelted bulk acting as the heat sink during the subsequent rapid solidification (Fig. 2.16) [2.37]. The resulting rapidly solidified material has the same composition as the underlying parent material, although the rapid solidification may produce a quite different microstructure and much improved properties. A second possibility is to preplace or inject alloy or dispersoid additions at the surface so that they are incorporated into the melt zone to form a surface region of composition different from the underlying bulk. The third possibility is to melt a different material preplaced on the surface so that

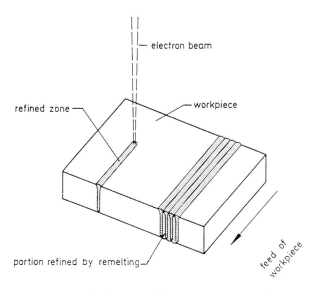

Fig. 2.16. Principle of rapid solidification at the surface of a block of material following rapid local melting with a traversing heat source [2.37]

mixing with the underlying material is limited to the minimum required to produce effective bonding. All three variants offer the practical possibility of generating a more durable surface on an underlying material that is in all other respects entirely adequate for the application in view. Both nanosecond and picosecond laser sources have been used to generate some quite spectacular non-equilibrium effects in surface melt zones as shallow as 0.1 µm in which cooling rates during solidification have been estimated to reach 10^{10} K/s [2.38] or more and solidification times to be as short as 10^{-9} s [2.39]. Both traversing laser and electron beams have been used to treat entire surfaces via repeated incremental displacement of the beam by the width of the melt zone at the start of each new traverse [2.37]. Samples with crack-free relatively smooth treated surfaces can now be produced by appropriate control of the process parameters. The technique can also be used to develop coupling between the traversing beam and the solidification front so that the effects on resulting microstructure of systematic variations in front velocity up to ≈ 1 m/s can be determined experimentally [2.40–42].

Important differences between surface melt traversing and droplet or spinning methods of rapid solidification are (*i*) that perfect thermal contact between the melt zone and the underlying solid is more or less assured, and (*ii*) that fluid flow plays a relatively insignificant part in determining the critical product dimensions (in this case, the depth z and width w of the melt zone). Making some simplifying assumptions, and under certain conditions, the depth melted (z_0) is expected [2.43, 44] to be governed by

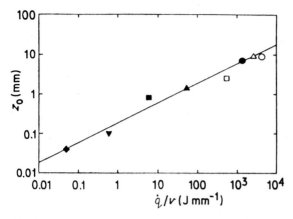

Fig. 2.17. Depth melted z_0 versus heat input per unit length $\dot{q}/v$ for TIG and electron beam traversing of type 310 stainless steel [2.44], data from [2.45]

$$z_o = 0.484\left(\frac{\dot{q}}{\rho c(T_F - T_A)v}\right)^{1/2}, \tag{2.11}$$

where $\dot{q}$ is the heat input per unit time from the heat source and v is the traverse velocity. Figure 2.17 shows conformity with this relationship for Tungsten Inert Gas (TIG) and electron beam traversing of type 310 stainless steel [2.44, 45]. The cooling rate in the path of a point source of heat moving at velocity v over the surface of a thick slab is [2.46]:

$$\dot{T} = 2\pi k v(T - T_A)^2/\dot{q}. \tag{2.12}$$

Eliminating $\dot{q}/v$ between (2.11) and (2.12) for $T = T_F$ thus gives

$$\dot{T} = 1.47k(T_F - T_A)/\rho c z_o^2, \tag{2.13}$$

which shows the same dependence of $\dot{T}$ on z_0 as (2.3) for ideal cooling of a slab.

A particular attribute of surface melt traversing is that, in the absence of significant bulk undercooling of the melt, the solidification front velocity V at any distance from the melt pool surface is given by $v\cos\theta$, where θ is the local inclination of the solidification front to the direction of travel of the heat source moving with velocity v, (Fig. 2.18) [2.47]. Thus, V increases from zero and normal to the free surface at the bottom of the melt pool to a value approaching v in magnitude and direction. When the product of solidification is microcellular or microduplex, etching of polished sections normal to the surface along the direction of traversing frequently reveals this progressive change in direction of solidification from $\theta = 90°$ initially to $\theta = \theta_m$ at the free surface (Fig. 2.19) [2.48]. Thus, in principle, the solidified melt zone contains the complete range of microstructures for the alloy for the range $0 < V < v\cos\theta_m$ although the initial increase in V is very rapid and so most of the melt zone solidifies at a velocity approaching $v\cos\theta_m$. Controlled front velocities V as high as several m/s at

2.4 Surface Melting Technologies   41

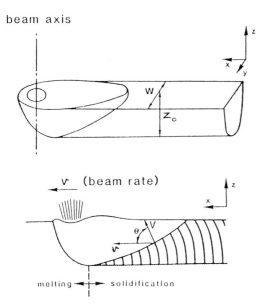

Fig. 2.18. Schematic of melt-pool in surface traverse melting showing position of heat source in relation to melting and solidification fronts [2.47]

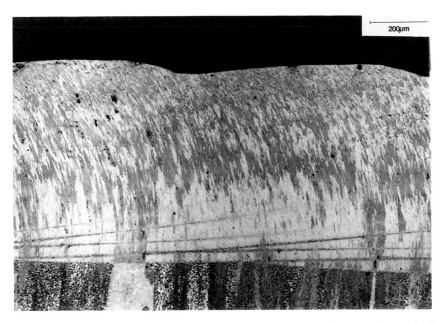

Fig. 2.19. Determination of θ_m from an etched longitudinal section of Al-33 wt.% Cu eutectic alloy TIG weld traversed at $v = 40$ mm/s [2.48]

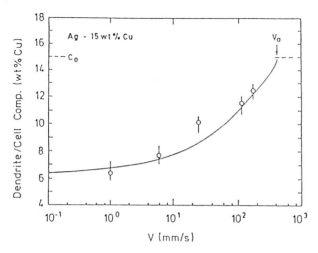

Fig. 2.20. Dendrite/cell composition versus growth velocity V for electron beam melt traversed Ag-15 wt.% Cu alloy. Points are measurements [2.49] and full line is predicted from dendrite growth theory. V_a represents the velocity for microsegregation-free solidification [2.50]

estimated temperature gradients G as high as 2000 K/mm have been obtained by such methods using a traversing laser as the heat source. Extended solute concentrations (Fig. 2.20) [2.49] and cell spacings (Fig. 2.21) [2.42] obtained by such methods are in good agreement with predictions [2.50, 51] of dendrite growth theory. Findings of such experiments can be combined with those of Bridgman growth studies at V up to a few mm/s and G up to ≈ 10 K/mm to generate solidification microstructure selection diagrams (Fig. 2.22) [2.52], in which fields of occurrence of particular microstructures are delineated on plots of V versus alloy concentration C_0. Conditions for one microstructure to be replaced by another can then be compared with predictions of competitive growth theory based on the supposition that the microstructure formed will be that which grows at the highest temperature at given V and C_0.

2.5 Consolidation Technologies

The purpose of consolidation as applied to rapidly solidified materials includes the following:

(*i*) to bond together particulate, ribbon or wire products,
(*ii*) to close isolated or linked porosity in the products,
(*iii*) to fragment and distribute surface films,
(*iv*) to generate a required section profile or three-dimensional shape with specified dimensions.

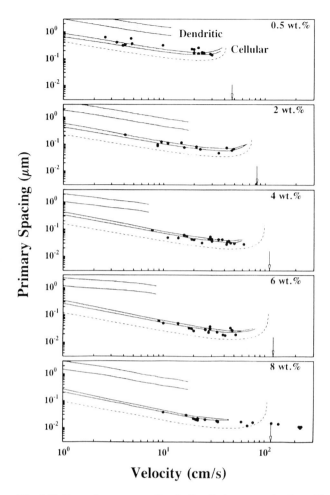

Fig. 2.21. Dependence on growth velocity of primary spacing for α-Al cells and dendrites for Al-0.5 to 8 wt.% Fe alloys. Shaded bands indicate predictions of an array model, broken lines are predictions from marginal stability at the tip and points are experimental data [2.42] from laser surface melt traversing [2.51]

Polymer bonding can be used in specific cases such as for producing small magnets from pulverized melt-spun Fe–Nd–B alloys. Melt-spun ribbons or wires can be suitably coated with a polymer or metal matrix and then warm or hot pressed to form a composite. A similar result can be obtained by pressing an assembly of alternate layers of ribbon or wire and matrix sheet, foil or powder, or by melt infiltration of a preform. It is evidently important to ensure that temperatures and times involved are not such as to promote degradation of the rapidly solidified material, as a result either of adverse chemical reaction with the matrix or of adverse transformation of the material itself.

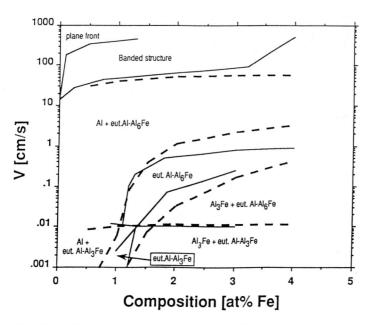

Fig. 2.22. Solidification microstructure selection diagram for Al-rich Al–Fe alloys Bridgman solidified at V up to 10 mm/s and laser surface melt traversed at V up to 5 m/s. Full lines are experimentally determined while broken lines are predictions of competitive growth modelling [2.52]

Closure of pores generated by the rapid solidification process itself can be achieved by a combination of cold and hot compaction on mechanical working or by hot compaction or working alone. These processes are also a very effective means of achieving direct interparticle bonding, especially where shear deformation is involved. Such shear deformation is essential if it is also necessary to fragment and distribute surface films. Production of a required section profile can be achieved by profiled rolling or extrusion of preconsolidated or encapsulated material, and three-dimensional shapes can be generated by cold or hot compacting into shaped dies or by forging of preconsolidated or encapsulated material. Once again, it is important to ensure that the thermomechanical excursions involved are not such as to generate adverse phase transformations or microstructural degradation of the rapidly solidified material.

Some special processes have been developed to address that particular issue. Dynamic compaction using explosives or a gas gun aims to produce rapid heating of the particle surfaces only to produce narrow melt zones. These then proceed to freeze and cool equally rapidly by heat flow to the unheated bulk of the particles to give an overall very small temperature rise while also achieving bonding and densification. Figure 2.23 shows an optical micrograph of an etched microsection of such a sample demonstrating that the melt zones formed in the compaction cycle were then solidified rapidly enough to generate metallic

Fig. 2.23. Optical micrograph of an etched microsection of shock-wave consolidated crystalline Ni-26at.%Mo-10at.%Cr-9at.%B alloy particulate made by pulverizing and annealing amorphous melt-spun material above its crystallization temperature. The white regions are amorphous material which form when the interparticle melt zones generated by the compaction cycle subsequently undergo rapid solidification by heat conduction into adjacent unheated material [2.53]

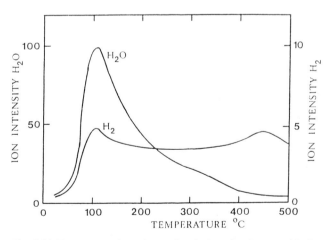

Fig. 2.24. Temperature dependence of evolution of moisture and hydrogen from nitrogen atomized Al-5.2wt.%Cr-1.6wt.%Zr-0.9wt.%Mn alloy powder according to quadrupole mass spectrometric analysis [2.55]

glasses in these regions [2.53]. Electrodischarge compaction has also been investigated as an alternative method of achieving rapid consolidation without adversely affecting the microstructure of the rapidly solidified material [2.54].

In addition to achieving bonding and densification in the consolidation process, it is very important in heat-treatable or high-temperature materials to ensure that the material is adequately degassed prior to consolidation. Failure to degas sufficiently can result in blistering or internal decohesion during subsequent heat treatment or service. This is a particular problem with alloys that require a high-temperature solution treatment before quenching and ageing. An example of the temperature dependence of the evolution of moisture and hydrogen from a gas-atomized aluminium alloy powder is shown in Fig. 2.24 [2.55]. Effective degassing can be achieved by vacuum treatment of the unconsolidated or partly consolidated material at a temperature higher than any it will experience subsequently. Vacuum hot pressing is an elegant means of combining the degassing and densification stages to provide a preform for further processing by rolling, extrusion or forging

References

2.1 P. Duwez: In *Techniques of Metals Research*, ed. by R.F. Bunshah (Interscience, New York 1968) Vol. 1, Pt. 1, pp. 343–358
2.2 H. Jones: *Treatise on Materials Science and Technology* **20**, 1–72 (Academic, New York 1981)
2.3 H. Jones: *Rapid Solidification of Metals and Alloys* (Institutions of Metallurgists, London 1982) Chaps. 1 and 2
2.4 H. Jones: In *Enhanced Properties in Structural Metals via Rapid Solidification*, ed. by F.H. Froes, S.J. Savage (ASM International, Metals Park, OH 1987) pp. 77–93
2.5 S.J. Savage, F.H. Froes: J. Met. **36**(2), 20–33 (1984)
2.6 T.R. Anantharaman, C. Suryanarayana: *Rapidly Solidified Metals: A Technological Overview*, (Trans. Tech., Aldermansdorf 1987) Chap. 2
2.7 C. Suryanarayana: in *Materials Science and Technology: A Comprehensive Treatment* **15**, 57–110 (VCH, Weinheim 1991)
2.8 R. Willnecker, P.M. Herlach, B. Feuerbacher: Phys. Rev. Lett. **62**, 2707–2710 (1989)
2.9 C.G. Levi, R. Mehrabian: Met. Trans. A **13**, 13–23 (1982)
2.10 W.J. Boettinger, L. Bendersky, J.G. Early: Met. Trans. A **17**, 781–790 (1986)
2.11 R.C. Ruhl: Mater. Sci. Eng. **2**, 314–319 (1968)
2.12 D.M. Herlach, F. Gillessen, T. Volkmann, M. Wollgarten, K. Urban: Phys. Rev. B **46**, 5203–5210 (1992)
2.13 U. Feurer, R. Wunderlin: Fachberichte Nr. 23 (DGM, Oberursel 1977)
2.14 H. Schmitt: Powd. Met. Int'l **11**, 17–21 (1979)
2.15 M. Ogushi, A. Inoue, H. Yamaguchi, T. Masumoto: Mater. Trans. JIM **31**, 1005–1010 (1990)
2.16 Anon: Adv. Mater. Proc. **135** (b), 12 (1989)
2.17 N. Dombrowski, W.R. Johns: Chem. Eng. Sci. **118**, 203 (1963)
2.18 H. Lubanska: J. Met. **22**(2), 45–49 (1970)
2.19 O.S. Nichiporenko, I. Naida: Sov. Powd. Met. Met. Ceram. **67**, 5099 (1968)
O.S. Nichiporenko: Sov. Powd. Met. Met. Ceram. **15**, 665 (1976)
2.20 B.P. Bewlay, B. Cantor: Met. Trans. B **2**, 866–912 (1990)
2.21 H.L. Liu, R.H. Rangel, E.J. Lavernia: Acta Mater. **42**, 3277–3289 (1994)

2.22 E.J. Lavernia, E.M. Gutierrez, J. Szekely, N.J. Grant: Int'l J. Rapid Solidificatiion **4**, 69–124 (1988)
2.23 B.A. Rickinson, F.A. Kirk, D.R.G. Davies: Powd. Met. **15**, 116–124 (1981)
2.24 D.H. Kirkwood: Int'l Mater. Rev. **39**, 173–189 (1994)
2.25 S. Annavarapu, D. Apelina, A. Lawley: Met. Trans. A **21**, 3237–3256 (1990)
2.26 K.J. Overshott: Electron. Power **25**, 347–350 (1979)
2.27 J. Edgington: In *Fibre-Reinforced Materials* (Inst. Civil Engineers, London 1977) pp. 129–140
2.28 I. Ohnaka: Int'l Rapid Solidification **1**, 219–236 (1985)
2.29 R.V. Raman, A.N. Patel, R.S. Carbonara: Proc. Powd. Met. **38**, 99–105 (1982)
2.30 A.L. Holbrook: Proc. Powd. Met. **41**, 679–684 (1986); Int'l J. Powd. Met. **22**, 39–45 (1986)
2.31 C. Gélinas, R. Angers, S. Pelletier: Mater. Lett. **6**, 359–361 (1988)
2.32 S. Pelletier, C. Gélinas, R. Angers: Int'l J. Powd. Met. **26**, 51–54 (1990)
2.33 S. Kavesh: In *Metallic Glasses* (Am. Soc. Met., Metals Park, OH 1978) pp. 36–73
2.34 P.H. Shingu, K.N. Ishihara: In *Rapidly Solidified Alloys*, ed. by H.H. Liebermann (Dekker, New York 1993) pp. 103–118
2.35 L.A. Anastiev; Mater. Sci. Eng. A **131**, 115–121 (1991)
2.36 O.P. Pandey, S.N. Ohja, G.M. Sarma, E.S. Dwarakadasa, T.R. Anantharaman: Indian J. Technol. **29**, 173–178 (1991)
2.37 B. Lux, W. Hiller: Prakt. Metallogr. **8**, 218–225 (1971)
2.38 M. von Allmen, M. Huber, A. Blatter, K. Affolter: Int'l J. Rapid Solidification **1**, 15–25 (1984)
2.39 F. Spaepen: In *Undercooled Alloy Phases*, ed. by E-W. Collings, C.C. Koch (TMS, Warrendale, PA 1987) pp. 187–205
2.40 W.J. Boettinger, D. Shechtman, R.J. Schaefer, F.S. Biancaniello: Met. Trans. A **15**, 55–66 (1984)
2.41 M. Zimmermann, M. Carrard, W. Kurz: Acta Met. **32**, 3305–3313 (1989)
2.42 M. Gremaud, M. Carrard, W. Kurz: Acta Met. Mater. **38**, 2587–2599 (1990)
2.43 N. Christensen, V. de L. Davies, and K. Gjermundsen: Brit, Weld. J. **12**, 54–75 (1965)
2.44 H. Jones: In *Rapid Solidification Processing: Principles and Technologies*, ed. by R. Mehrabian, B.H. Kear, M. Cohen (Claitor's, Baton Rouge, LA 1978) pp. 28–45
2.45 Y. Arata, F. Matsuda, K. Nakata: Trans. Jpn. Weld. Inst. **5**(1), 47–52 (1976)
2.46 C.M. Adams: Weld. J. Res. Suppl. **37**, 210s–215s (1958)
2.47 M. Rappaz, M. Gremaud, R. Dekumbis, W. Kurz: In *Laser Treatment of Materials*, ed. by B.L. Mordike (DGM, Oberursel 1987) pp. 43–53
2.48 S.A. Moir, H. Jones: J. Mater. Sci. Lett. **10**, 1199–1201 (1991)
2.49 W.J. Boettinger, L.A. Bendersky, S.R. Coriell, R.J. Schaefer, F.S. Biancaniello: J. Cryst. Growth **80**, 17 (1987)
2.50 W. Kurz, R. Giovanola, R. Trivedi: J. Cryst. Growth **91**, 123–125 (1988)
2.51 J.D. Hunt, S.-Z. Lu: Mater. Sci. Eng. A **173**, 79–83 (1993)
2.52 W. Kurz, P. Gilgien: Mater. Sci. Eng. A **178**, 171–178 (1994)
2.53 N.N. Thadhani, T. Vreeland Jr.: Acta Met. **34**, 2323–2334 (1986)
2.54 H.-R. Pak, D.K. Kim, K. Okazaki: Mod. Dev. Powd. Met. ed by P.U. Gummeson and D.A. Gustafson (APMI, Princeton, NJ 1988) **19**, pp. 591–602
2.55 M.M. Silva, H. Jones, C.M. Sellars: Proc. PM'90 (Inst. of Metals, London 1990) Vol. 2, pp. 315–318

3 Structure and Characterization of Rapidly Solidified Alloys

C.N.J. Wagner, M.A. Otooni, and W. Krakow

To understand the most basic properties of the amorphous solids, being mechanical, chemical electrical or magnetic, a full description of the atomic structure in these materials is required. However, the determination of the physical properties of amorphous solids or of liquids without a definite atomic structure has remained a formidable task to accomplish. To circumvent this difficulty, the prediction of the physico-chemical properties has been largely based on the principles of thermodynamics of a randomly distributed collection of atoms [3.1–4]. For example, from the application of the pair distribution function (3.17), the total energy of a system, which is largely dependent upon a two-body correlation [3.5], can be calculated. Along the same line, more recently a more rigorous attempt to determine the physical properties of an amorphous solid has been made by using the concept of local structural fluctuations and its associated local structural parameters [3.6–7]. By application of this concept, a method for calculating the energy of the system has been made. In this methodology the principles of elastic approximation have been described in terms of the local atomic-level stresses and strains.

There are two important features of this theoretical approach. First, it provides a unique and mathematically suitable technique for calculating local stresses and elastic constants of an amorphous material. These data can then be employed to develop thermomechanical stability criteria which are based on the local excess energies of the local fluctuations. Second, it reveals the nature of the moduli distribution associated with these local distributions.

Based on these atomic-level observations, plausible explanations have been provided to bring about fundamental understanding of the enhanced strength, ductility, wear resistance and corrosion, as well as electrical, and magnetic properties of the rapidly solidified materials.

3.1 Characterization Techniques

3.1.1 Structural Characterization

There are several experimental procedures which can be used in characterizing and assessing the structure of rapidly solidified alloys. These include X-ray and neutron diffraction, scanning and Transmission Electron Microscopy (TEM) electric and magnetic measurements, and differential scanning calorimetry.

X-Ray Radial Distribution Function (XRDF) studies are usually performed to characterize the amorphous nature of the rapidly solidified material. The diffraction capability of transmission electron microscopy is useful in disclosing the diffused diffraction rings, which is indicative of the amorphous state. Also, the image capability of the TEM could be used to provide the bright-field image having salt-and-pepper contrast of the amorphous materials. The capability of high resolution of TEM is very useful in showing the degree of atomic structural or chemical order. The application of differential scanning calorimetry is useful to shed light on the extent of undercooling (quench), determination of glass-transition temperature, transformation, viscosity fluctuation, and the annealing history of the quenched state [3.8–14]. Standard metallurgical techniques are principally employed for characterization of the mechanical properties such as strength, ductility, wear resistance, hardness, and grain-size measurements [3.15–21].

3.1.2 X-Ray Radial Distribution Function

In this approach, X-ray studies are performed to calculate the reduced total interference function $i(q) = I(q) - 1$ (3.15) and the pair distribution function $G(r)$ (3.17) according to equation

$$G(r) = 4\pi r [\rho(r) - \rho_0] = (2/\pi) \int q[i(q)] \sin(qr)\, dq,$$

where $\rho(r)$ is the total radial density or atomic distribution function (3.22) and ρ_0 is the average density of the atoms. $G(r)$ is also called the total reduced atomic distribution function (3.17). The quantities $q \equiv Q$ (3.4) and r are the length of the wave vector (3.4) and the atomic separation, respectively [3.22].

XRDF provides two important and readily obtainable kinds of information. First, a plot of $\rho(r)$ or $G(r)$, shown in Fig. 3.1 for amorphous and nanocrystalline Fe–W alloys, reveals a unique, harmonic graph with a series of broad peaks which attenuate in amplitude with increasing distance r. Secondly, the distances between first nearest neigbour atoms in the amorphous structure can be readily calculated from these plots. For example, in a sample of amorphous Cu–Zr used in the radial distribution function studies, the atomic separation between two dissimilar atoms, Cu and Zr, was determined to be 2.72Å and the Zr to Zr separation was found to be 3.15Å (Table 3.1). From these atomic separations and the similarities between detailed features of $G(r)$ and those of other amorphous alloys and liquids, the glassy nature of the alloy can be characterized [3.16].

Although results from the XRDF analyses could be useful in indicating whether or not materials under investigation are amorphous, they could not totally exclude the possibility of the presence of chemical short-range order, which is only detectable by more rigorous experimental procedures. In fact, more recent studies have shown the presence of chemical (compositional) short-range regions in almost all glass–metal alloys and have also established a firm thermodynamic basis to explain their influence upon properties of amorphous solids [3.8]. The quantitative resolution of the XRDF technique to delineate

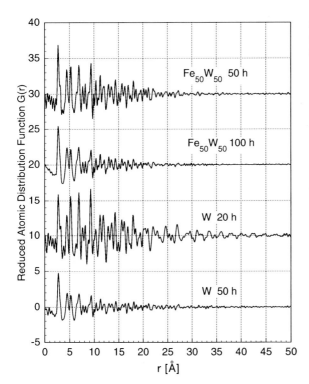

Fig. 3.1. X-ray reduced atomic distribution functions $G_l(r)$ of W and $Fe_{50}W_{50}$ alloys

ordered regions comprising less than 1–2% of the host amorphous matrix is, however, severely limited. Unless special experimental procedures are employed, short-range-order regions may remain experimentally undetected.

In amorphous samples with little or no regions of short-range order, both reduced interference function $i(q)$ and reduced radial distribution $G(r)$ are generally smooth. The presence of a kink in the second peaks are generally indicative of short-range-order regions within an amorphous matrix. The presence of these short-range orders will have dramatic effects on the magnetic properties of materials processed by rapid solidification and will be borne out by Mössbauer and other magnetic and electrical experimental procedures such as magneto-

Table 3.1. XRDF – peak positions of $i(q) Å^{-1}$ and $G(r) Å$ for Cu_{60}–Zr_{40} and Ni_{60}–Nb_{40} amorphous alloys [3.9]

Alloy							
Cu_{60}–Zr_{40}	$i(q) Å^{-1}$	2.82	–	4.82	5.33	–	7.67
	$G(r) Å$	2.72	3.15	4.75	5.20	5.70	7.20
Ni_{60}–Nb_{40}	$i(q) Å^{-1}$	2.94	–	4.95	5.84	–	7.55
	$G(r) Å$	2.65	–	4.53	5.05	6.78	–

resistance and Hall-effect measurements. For example, recently, it has been proposed that the transport properties of magnetically ordered amorphous alloys can contain two contributions of nearly equal magnitude, one arising directly from the structural order and another from topological order which is impressed on the spin lattice, i.e., an indirect effect of the structural order. The application of such experimental tools as Mössbauer, magneto-resistance and Hall-effect measurements will have distinct advantages.

3.1.3 High-Resolution Electron Microscopy

High-resolution electron microscopy (image and diffraction) can be considered as a powerful tool in the investigation of amorphous, ultra-fine grain and rapidly processed solidified materials. The bright-field image of the as-formed glassy alloys is usually characterized by having no characteristic feature other than a salt-and-pepper contrast (Fig. 3.2) arising from local variation of either thickness or composition, or amorphous structure. The latter is due to a non-uniform cooling history. The diffraction pattern of the amorphous state is usually fea-

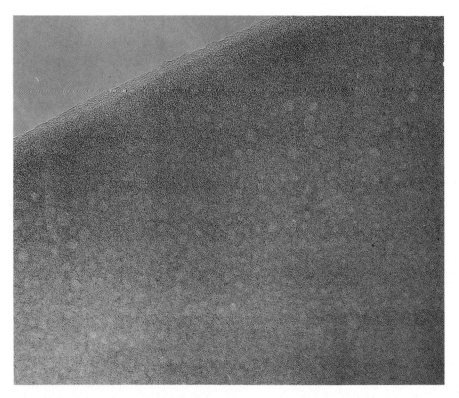

Fig. 3.2. Showing the bright-field electron micrograph of an amorphous Cu–Zr alloy with salt-pepper contrast

Fig. 3.3. Nature of the diffused diffraction rings from an originally amorphous Cu–Zr alloy

tureless (Fig. 3.3). It is characterized by broad diffused rings, indicating the random structure typically obtained from liquids or vitreous materials such as amorphous quartz (tridimite or crystabolite). An ordinary electron microscope will be sufficient to obtain such preliminary information. The application of high-resolution electron microscopy will have dramatic influence on deciphering such a phenomenon as clustering at the early stage of crystallization, the presence of unambiguous order in RS-processed materials based on the resolution of the lattice, and pin-pointing of the appropriate atoms in respective sites and their verification in comparison with simulation models, growth characteristics during annealing and aging, presence of micro cracks and voids, crack coalescence, visco-elastic information and many other relevant information. By using high-resolution techniques, one can assess the onset of nucleation and any nucleation density fluctuation which could arise from short-range order (compositional and/or topological). Similarly, from high-resolution electron microscopy one can assess regions of high or low atomic stress and strain within an amorphous matrix or, in the case of magnetic materials, decipher regions of different spin orientations (domains). For the most part, the differences in contrast in bright-field electron micrograph images of the amorphous materials is caused by the variation in thickness of the specimen. Also, on a finer scale, there are some cluster regions within the amorphous matrix which have a well-developed crystal lattice and could only be resolved by using more advanced techniques of electron microscopy (Figs. 3.4–7).

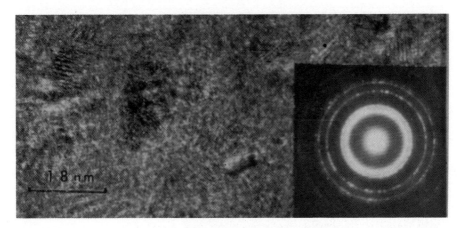

Fig. 3.4. Nature of crystallization within an amorphous matrix. Note the nature of localized crystallization possibly influenced by the thickness of the ribbon at the time of processing

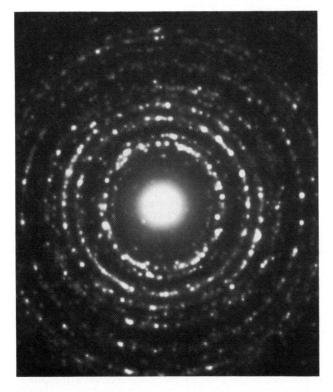

Fig. 3.5. Nature of the electron diffraction pattern with the associated diffraction rings and superposed diffraction spots of Cu–Zr alloy material

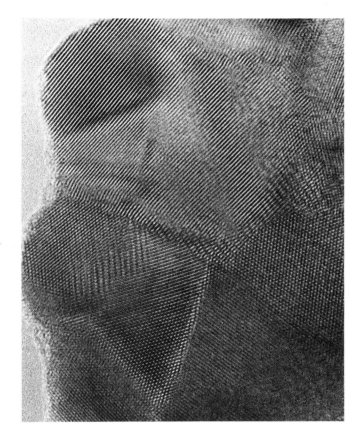

Fig. 3.6. Development of lattice fringes from an originally amorphous specimen of Cu–Zr alloy. Note the untransformed region of the specimen

3.1.4 Differential Scanning Calorimetry – Phase Transformation and Separation

Differential Scanning Calorimetry (DSC) is most extensively used for detecting the onset of crystallization from the amorphous state. Figures 3.8, 9 depict several DSC tracings during crystallization of the Cu–Zr glass upon isothermal annealing. The glass-transition temperature as well as transformation temperature can be obtained from the DSC tracing.

Another aspect of the amorphous-crystalline transition in metal glasses which deserves critical examination is the phenomenon of phase separation.

Since, during the rapid solidification process extended solid solution occur, the subsequent annealing results in the precipitation of the excess solute, most commonly as a single phase. This single phase, known as fine-scale product, adversely affects the mechanical properties of the rapidly solidified materials. Figures 3.7a–c shows high-resolution electron micrograph of an isothermally annealed sample of an initially amorphous Cu_{60}–Zr_{40} alloy. The electron

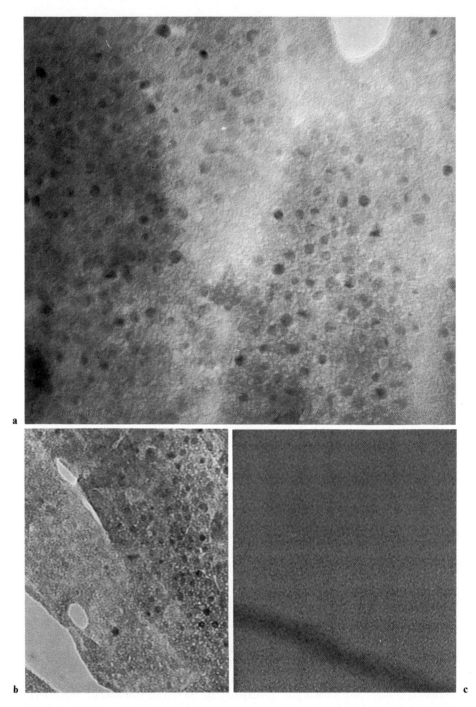

Fig. 3.7a–c. Nature of oxides distribution after annealing in air of an initially amorphous Cu–Zr alloy materials: (**a**) Incomplete coverage due to short time interval of annealing; (**b**) complete coverage of the surface due to the longer time of annealing

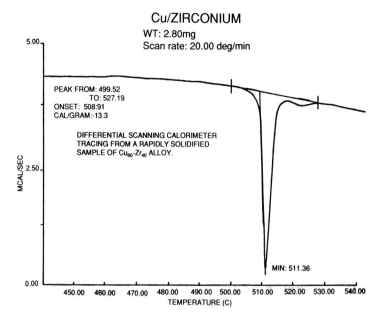

Fig. 3.8. Typical differential scanning calorimetric trace from a specimen of Cu–Zr alloy annealed to 480 °C

micrograph shows resolved lattice lines of the (200) planes of copper with the interplanar spacings of 1.77 Å. There have been adequate thermodynamic explanations for the occurrence of this fine-scale precipitation and how it contributes to the thermodynamic instabilities of metal glasses in general [3.8].

3.1.5 Electrical Resistivity

The Electrical-Resistivity (ER) measurement is also a useful tool in characterizing rapidly solidified alloys. One notable feature of this technique is in detecting the transition from the amorphous to the crystalline state. This transition is generally abrupt and is indicated by a sudden increase in the resistivity (positive coefficient of resistivity) (Fig. 3.10). The increase of the ER is due to the change in the electrical transport phenomenon due to early formed nuclei. Evidently, the conduction mechanism will be drastically altered during subsequent growth from the amorphous state. The ER technique is a sufficiently powerful tool to allow detection of the early stages of the transformation, and to delineate steps in the kinetics of transformation in which the electron conduction mechanism is strongly affected by the size and distribution of the early nuclei [3.8–10]. More recently, attempts have been made to relate the glass-forming ability with the electronic properties in the amorphous or liquid state. More specifically, it has been suggested that alloys with a negative or small

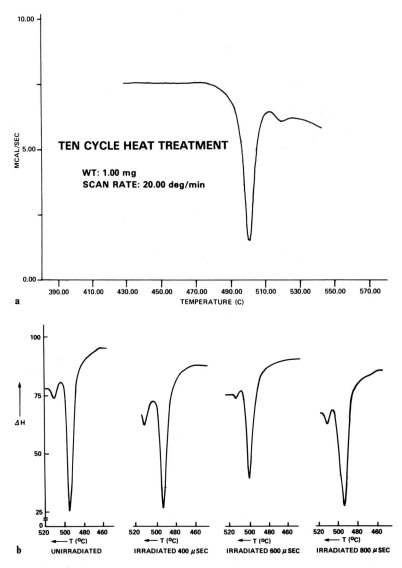

Fig. 3.9a,b. Differential scanning calorimetric traces from a series of Cu–Zr alloy cyclically annealed. (a) Annealed one times; (b) annealed successively

positive temperature dependence of electrical resistivity are expected to possess an improved stability of the amorphous phase. For most alloys, the trend in electrical resistivity behaviour is similar. There is a relatively small positive temperature coefficient of resistivity in the amorphous state, followed by a sharp decrease as the crystallization occurs. The onset of this decrease has been taken as the crystallization temperature T, which still depends on the heating rate during the annealing process (Figs. 3.11–12). An independent technique for measurement of the transformation temperature is often necessary to ensure the

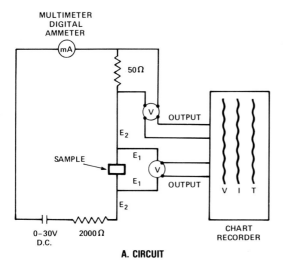

Fig. 3.10. Schematic diagram showing a typical experimental apparatus for electrical measurements

results from ER measurements. In the case of rapidly solidified magnetic alloys, an independent determination of T can be made from thermomagnetic measurements. In general, resistivity measurements can be made under cryogenic environment, 1.5–50 K, or room-temperature environment. The aim of the low-temperature-resistivity measurement is to investigate more closely the correlation between structural changes and the resistance behaviour, i.e., to minimize the thermal contribution to resistivity.

The low-temperature-resistance results are particularly sensitive to structural changes of the sample and can provide a suitable way of monitoring the morphology of phase transformations. Absolute resistivity values can be obtained from the geometrical shape factor deduced from the measured mass, the density and the length of the sample.

In some instances in which problems concerning atomic site occupations and charge-density measurements may be **necessary**, Hall effect measurements have been performed. These measurements **basically assume a most widely accepted nearly free** electron approach to the electron structure of amorphous alloys. The relationship between RH, the Hall Resistance (RH) and n, the volume concentration of the free electrons, and e, the electron charge is expressed as

$$RH = (1/n)e.$$

The information obtained by this measurement is, however, of little use for the process metallurgy. Quantitative information could be, however, obtained from the Hall data on samples which have been step-wise annealed. It has been found that the absolute value of the RH decreases with increasing temperature of annealing. When fully crystallized, the alloy may have smaller values of RH compared to those of the amorphous state. This could be explained by the formation of bonding levels lying deeply below the Fermi level in the band structure.

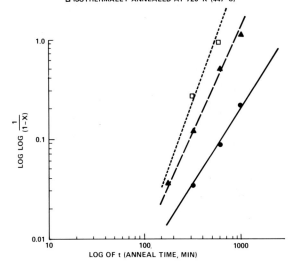

Fig. 3.11. Typical log-log plot of the percent X transformed of the originally amorphous alloy as a function of $\log t$ (t = time in min)

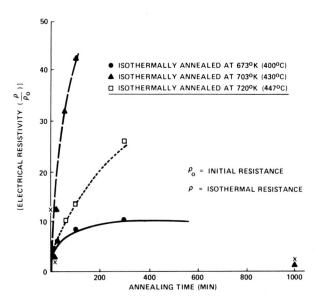

Fig. 3.12. Schematic curves showing the variation of the electric resistivity of isothermally annealed specimens

3.1.6 Microhardness Measurements

Microhardness measurements are generally made by using a Vickers pyramid indenter on various micro-hardness testers, equipped with a low load of either 25 or 50 g to insure that there is no possibility of stress relaxation close to the specimen edges. Indentation can be made directly on a flat disc surface or on mounted and polished through-thickness cross sections. In the isothermally annealed specimens of the majority of the rapidly solidified alloys, the microhardness initially increases with the annealing temperature. This trend will subsequently change to a decrease of the microhardness. The abrupt change of the microhardness at the early stage of crystallization is generally interpreted in terms of atomic relaxation which is often expected from fast cooling molten alloys. If it is assumed that the amorphous alloy is made of "frozen-in" nuclei formed during the rapid solidification, it is likely that the supply of thermal energy during annealing will facilitate atomic rearrangements and will actually lead to the process of cluster formation.

The sharp, early rise in microhardness, in essence, signals the onset of nucleation from an amorphous matrix, similar to the increase of hardness observed during the well-known precipitation hardening processes of polycrystalline supersaturated alloys. In the case of supersaturated alloys, the nuclei formed during precipitation will impede dislocation motion by acting as a barrier. The climb of dislocations over the barrier assisted by the supplied thermal energy would cause the subsequent softening of the alloy. In the case of the amorphous alloy, it is possible to have crystallized ordered regions having dislocation networks. The onset of the dislocation network formation may be interpreted as a mechanism involved in the sharp increase of microhardness. In some alloys, the sharp, early increase in the microhardness may be due to segregation and/or appearance of a new phase whose proportion will increase during subsequent annealing stage [3.14, 20, 21].

3.1.7 Mössbauer Spectroscopy

This technique is used for the characterization of the magnetic glassy alloys such as Fe–B and others having magnetic properties. In most applications, a conventional constant acceleration spectrometer equipped with a multichannel analyzer is used. The obtained spectra contain about 5×10^3 counts per channel. The spectra consist of two parts: *i*) An overlap-free sub spectrum and *ii*) a broad and somewhat structureless spectrum. A new technique has been developed to separate these two spectral lines and to draw conclusions on the probability distribution of hyperfine fields. The broad and featureless spectrum is indicative of the glass structure. Its composition can be determined by fitting the spectrum with an average hyperfine field and the isomer shift. The technique can be successfully employed to predict the structure due to the crystallization, and to verify the composition of the phase(s) formed after annealing. The technique is sufficiently refined to allow determination of the atomic environment, nearest atomic sites occupancy (for similar and dissimilar atoms) as well as determination of the smaller to the greater hyperfine field sites.

Some traces of thermal relaxation in the glassy state can be detected by static magnetic measurements. At a temperature close to the transition temperature a change in magnetization occurs. The magnetization decreases rapidly with increasing temperature. The Curie point can be determined from the sharp kink-point. An indication of the extent of thermal relaxation can be inferred from the general shape of the kink. The higher extent of relaxation is indicated by the roundness of the kink's tip.

3.2 Total Scattering Intensity from Amorphous and Nanocrystalline Alloys

Neutron, X-ray and electron diffraction provide powerful techniques to elucidate the structure of amorphous and nanocrystalline materials [3.22], prepared by rapid quenching from the liquid state or by mechanical alloying in high-energy ball-mills. In any scattering or diffraction experiment on poly- or non-crystalline samples, exhibiting random orientation of their units of structure, one measures the scattered neutron, electron or X-ray intensity as a function of the length Q of the diffraction vector $\mathbf{Q} = \mathbf{k}_1 - \mathbf{k}_0$, where $\mathbf{k}_0$ is the wave-vector of the incident beam of energy E_0 and $\mathbf{k}_1$ is the wave-vector of the scattered beam of energy E_1. The angle 2θ between the indicent beam $\mathbf{k}_0$ and the scattered or diffracted beam $\mathbf{k}_1$ is called the scattering or diffraction angle.

In the scattering process, we distinguish two cases: (i) elastic scattering where the incident and scattered beams have equal energy, i.e., $E_1 = E_0$, and (ii) inelastic scattering in which measures are taken to define both incoming and scattered energies. In this chapter, we will discuss only the elastic scattering of neutrons, electrons and X-rays, i.e., we will employ the *static approximation of scattering*. Then $k_0 = k_1 = 2\pi/\lambda$, and the wavelength λ is related to the energies of the neutrons, electrons and X-rays by:

$$E_{\text{neutron}} = [h^2/(2m_n)]/\lambda^2 = 81.787/\lambda^2, \tag{3.1}$$

$$E_{\text{electron}} = [h^2/(2m_e)]/\lambda^2 = 150/\lambda^2, \tag{3.2}$$

$$E_{\text{X-ray}} = hc/\lambda = 12.40/\lambda, \tag{3.3}$$

respectively, where h is Planck's constant, m_n is the mass of the neutron, m_e is the mass of the electron, and c is the velocity of the photon. In (3.1–3), we express the energy E_{neutron} of the neutron in meV, the energy E_{electron} of the electron in eV, the energy $E_{\text{X-ray}}$ ray of the X-ray photon in keV, and the wavelength λ in Å. Thus, the length Q of the diffraction vector $\mathbf{Q}$ can be expressed as [3.22]:

$$Q = 4\pi(\sin\theta)/\lambda \tag{3.4}$$

$$= 1.0134(\sin\theta)E_{\text{X-ray}}$$

$$= 1.026(\sin\theta)(E_{\text{electron}})^{1/2}$$

$$= 1.3895(\sin\theta)(E_{\text{neutron}})^{1/2}. \tag{3.5}$$

3.2.1 Atomic Distribution Functions

The atomic short-range order in binary amorphous or nanocrystalline alloys, consisting of elements of types 1 and 2, is characterized by three atomic pair distribution functions $\rho_{ij}(r)$ describing the number of j-type atoms per unit volume at the distance r about an i-type atom [3.23, 24] i.e., $\rho_{11}(r)$, $\rho_{22}(r)$ and $\rho_{12}(r)$, because $\rho_{21}(r)/c_1 = \rho_{12}(r)/c_2$, c_i being the atomic concentration of element i. An alternative description of the atomic short-range order is given by the number–concentration correlation functions $\rho_{nc}(r)$ [3.25]. The number–number distribution function $\rho_{nn}(r)$ represents the topological short-range order, the concentration–concentration correlation function $\rho_{cc}(r)$ describes the chemical short-range order, and the number–concentration correlation function $\rho_{nc}(r)$ is related to the size difference between atoms 1 and 2 [3.26]. The functions $\rho_{ij}(r)$ and $\rho_{nc}(r)$ are linearly related, i.e.,

$$\rho_{nn}(r) = c_1\rho_{11}(r) + c_2\rho_{22}(r) + 2c_1\rho_{12}(r) = c_1\rho_1(r) + c_2\rho_2(r), \tag{3.6}$$

$$\rho_{cc}(r) = c_2\rho_{11}(r) + c_1\rho_{22}(r) - 2c_1\rho_{12}(r) = [c_2\rho_1(r) + c_1\rho_2(r)]\alpha(r), \tag{3.7}$$

$$\rho_{nc}(r) = c_1c_2\{[\rho_{11}(r) + \rho_{12}(r)] - [\rho_{22}(r) + \rho_{21}(r)]\}$$

$$= c_1c_2[\rho_1(r) - \rho_2(r)], \tag{3.8}$$

where $\rho_1(r) = \rho_{11}(r) + \rho_{12}(r)$, $\rho_2(r) = \rho_{21}(r) + \rho_{22}(r)$ and $\alpha(r)$ is the generalized Warren short-range order parameter [3.24] defined as

$$\alpha(r) = 1 - \rho_{12}(r)/\{c_2[c_2\rho_1(r) + c_1\rho_2(r)]\}. \tag{3.9}$$

3.2.2 Scattered Intensity

We define the scattered intensity $S_d(Q)$ by

$$S_d(Q) = I_a(Q) - [\langle f^2 \rangle + \langle f \rangle^2 \rho_0 V(Q)], \tag{3.10}$$

where $I_a(Q)$ is the scattered intensity per atom, $\langle f^2 \rangle = c_1(f_1)^2 + c_2(f_2)^2$ is the mean-square scattering factor, $\langle f \rangle = c_1 f_1 + c_2 f_2$ is the mean scattering amplitude, and ρ_0 is the average atomic density. The term $\langle f \rangle^2 \rho_0 V(Q)$ represents the small-angle scattering, and $V(Q)$ is the Fourier transform of the size function $v(r)$, i.e.,

$$v(r) = (1/V) \int s(u)s(u+r)\,du, \tag{3.11}$$

where $s(r) = 1$ inside the volume V of the particle and zero outside [3.22, 23].

The intensity per atom $S_d(Q)$ can be expressed as the weighted sum of the partial interference functions or structure factors $I_{ij}(Q)$, i.e.,

$$S_d(Q) = (c_1 f_1)^2 [I_{11}(Q) - 1] + (c_2 f_2)^2 [I_{22}(Q) - 1]$$
$$+ 2c_1 f_1 c_2 f_2 [I_{12}(Q) - 1]. \tag{3.12}$$

The Fourier transforms of $Q[I_{ij}(Q) - 1]$ yield the partial reduced atomic distribution functions $G_{ij}(r) = 4\pi r\{[\rho_{ij}(r)/c_j] - \rho_0\}$, i.e.,

$$G_{ij}(r) = 4\pi r\{[\rho_{ij}(r)/c_j] - \rho_0\} = (2/\pi) \int Q[I_{ij}(Q) - 1](\sin Qr)\,dQ. \tag{3.13}$$

The scattered intensity $S_d(Q)$ can also be expressed as the weighted sum of the partial structure factors $S_{nc}(Q)$ which are related to the Fourier transforms of the partial correlation functions $\rho_{nc}(r)$, i.e.,

$$S_d(Q) = \langle f \rangle^2 [S_{nn}(Q) - 1] + (\langle f^2 \rangle - \langle f \rangle^2)\{[S_{cc}(Q)/(c_1 c_2)] - 1\}$$
$$+ 2\Delta f \langle f \rangle S_{nc}(Q), \tag{3.14}$$

where $\Delta f = f_1 - f_2$. In cases where $\langle f \rangle = 0$, $S_d(Q)$ yields directly the concentration structure factor $S_{cc}(Q)$. On the other hand, when $\Delta f = 0$, $S_d(Q)$ is equal to the number density structure factor $S_{nn}(Q)$. The Fourier transforms of $Q[S_{nn}(Q) - 1]$, $Q\{[S_{cc}(Q)/(c_1 c_2)] - 1\}$ and $QS_{nc}(Q)$, yield the number-concentration correlation functions $4\pi r[\rho_{nn}(r) - \rho_0]$, $4\pi r\rho_{cc}(r)$ and $4\pi r\rho_{nc}(r)$, respectively.

In general, the small-angle scattering $\langle f \rangle^2 \rho_0 V(Q)$ is not observable in amorphous metals, or is subtracted from the scattered intensity from nanocrystalline metals. For large values of Q, $I_a(Q) = \langle f^2 \rangle$ and $S_d(Q) = 0$. Since $f_i(Q)$ is a decreasing function with increasing Q for X-rays and electrons, it is customary to introduce the total interference function $I(Q)$ and the total structure factor $S(Q)$, defined as:

$$I(Q) = [S_d(Q)/\langle f \rangle^2] + 1 = \{I_a(Q) - [\langle f^2 \rangle - \langle f \rangle^2]\}/\langle f \rangle^2 \tag{3.15}$$

$$S(Q) = [S_d(Q)/\langle f^2 \rangle] + 1 = I_a(Q)/\langle f^2 \rangle \tag{3.16}$$

The definition of $S(Q)$ has the advantage that it is valid even when $\langle f \rangle = 0$ as is the case for a "zero alloy", an alloy containing the appropriate amount of an element or isotope with negative scattering length for neutrons. Examples of $I(Q)$ and $S(Q)$ are shown in Fig. 3.13 for amorphous $Be_{38}Ti_{62}$ [3.27] and nanocrystalline Fe [3.28].

3.2.3 Reduced Atomic Distribution Functions

The total interference function $I(Q)$ or structure factor $S(Q)$ permit us to evaluate the total reduced atomic distribution functions $G_I(r)$ or $G_S(r)$, i.e.,

3.2 Total Scattering Intensity from Amorphous and Nanocrystalline Alloys   65

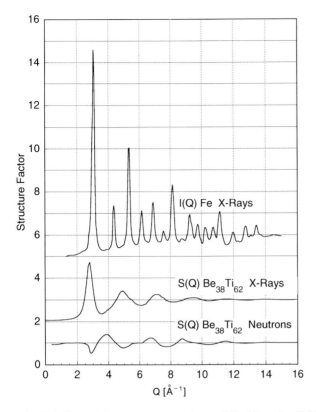

Fig. 3.13. X-ray and neutron structure factor $S(Q)$ of $Be_{38}Ti_{62}$ [3.27] and X-ray interference function $I(Q)$ of nanocrystalline Fe [3.28]

$$G_I(r) = (2/\pi) \int Q[I(Q) - 1]M(Q) \sin Qr \, dQ$$
$$= 4\pi r[\rho(r) - \rho_0]v(r)$$
$$= \sum_i \sum_j (c_i f_i c_j f_j / \langle f \rangle^2) G_{ij}(r) v(r), \tag{3.17}$$

$$G_S(r) = (2/\pi) \int Q[S(Q) - 1]M(Q) \sin Qr \, dQ$$
$$= 4\pi r[\rho(r) - \rho_0](\langle f \rangle^2 / \langle f^2 \rangle) v(r)$$
$$= \sum_i \sum_j (c_i f_i c_j f_j / \langle f^2 \rangle) G_{ij}(r) v(r), \tag{3.18}$$

$$G_S(r) = 4\pi r\{(\langle f \rangle^2 / \langle f^2 \rangle)[\rho_{nn}(r) - \rho_0] + \{[\langle f^2 \rangle - \langle f \rangle^2]/\langle f^2 \rangle\}\rho_{cc}(r)$$
$$+ 2[\Delta f \langle f \rangle \langle f^2 \rangle]\rho_{nc}(r)\}v(r), \tag{3.19}$$

where $v(r)$ is defined in (3.11), $M(Q)$ is a modification function [3.23, 29, 30]

$$M(Q) = \frac{[\sin \pi(Q/Q_{max})]}{[\pi Q/Q_{max}]}, \tag{3.20}$$

or

$$M(Q) = \exp(-\alpha Q^2) \quad \text{for } Q < Q_{\max},$$
$$= 0 \quad \text{for } Q > Q_{\max}; \quad (3.21)$$

$Q_{\max}$ being the maximum value of the Q-range and α an artificial temperature factor, e.g., $\alpha = 0.005$, and $\rho(r)$ is the total atomic pair distribution function, i.e.,

$$\rho(r) = (c_i f_i c_j f_j / \langle f \rangle^2) \rho_{ij}(r)/c_j. \quad (3.22)$$

The total differential correlation function $D(r)$ is given by [3.24]

$$\begin{aligned}
D(r) &= (2/\pi) \int Q S_d(Q) M(Q) \sin Qr \, dQ \\
&= \langle f^2 \rangle G_S(r) \\
&= \langle f \rangle^2 G_1(r) \\
&= 4\pi r [\rho(r) - \rho_0] \langle f \rangle^2 v(r) \\
&= \sum_i \sum_j 4\pi r [c_i f_i f_j \rho_{ij}(r) - \langle f \rangle^2 \rho_0] v(r) \\
&= 4\pi r \{\langle f \rangle^2 [\rho_{nn}(r) - \rho_0] \\
&\quad + [\langle f^2 \rangle - \langle f \rangle^2] \rho_{cc}(r) + 2\Delta f \langle f \rangle \rho_{nc}(r)\} v(r). \quad (3.23)
\end{aligned}$$

The total correlation function $T(r)$ can be deduced from $D(r)$, i.e.,

$$\begin{aligned}
T(r) &= 4\pi r \rho(r) \langle f \rangle^2 v(r) \\
&= 4\pi r \sum_i \sum_j c_i f_i f_j \rho_{ij}(r) v(r) \\
&= D(r) + 4\pi r \rho_0 \langle f \rangle^2 v(r) \\
&= 4\pi r \{\langle f \rangle^2 \rho_{nn}(r) + [\langle f^2 \rangle - \langle f \rangle^2] \rho_{cc}(r) + 2\Delta f \langle f \rangle \rho_{nc}(r)\} v(r). \quad (3.24)
\end{aligned}$$

The radial distribution function $R(r)$ describes the number of atoms in the spherical shell of radius r and unity thickness, i.e.,

$$\begin{aligned}
R(r) &= 4\pi r^2 \rho(r) v(r) \\
&= 4\pi r^2 \rho_0 v(r) + r G_1(r) \\
&= r T(r)/\langle f \rangle^2. \quad (3.25)
\end{aligned}$$

In amorphous or liquid samples with sizes larger than micrometers it can be assumed that $v(r) = 1$ over the range of r where $\rho(r) - \rho_0$ is different from zero. This is not true for nanocrystalline or nano-amorphous particles with dimensions of one to three nanometers.

3.2.4 Coordination Numbers in Binary Amorphous Alloys

Since the reduced atomic distribution function $G(r)$ of amorphous materials exhibits only few well-defined peaks, it is customary to concentrate on the first peak which is used to calculate the total coordination numbers N_I and N_S, i.e., [3.31]

$$N_I = \int 4\pi r^2 \rho(r)\,dr = \int rG_I(r)\,dr + 4\pi r^2 \rho_0$$
$$= \sum_i \sum_j [c_i f_i c_j f_j / \langle f \rangle^2] N_{ij}(r)/c_j, \tag{3.26}$$
$$N_S = \int 4\pi r^2 \rho(r) \langle f \rangle^2 / \langle f^2 \rangle\,dr = \int rG_S(r)\,dr + 4\pi r^2 \rho_0 \langle f \rangle^2 / \langle f^2 \rangle$$
$$= [\langle f \rangle^2 / \langle f^2 \rangle] N_{nn} + [(\langle f^2 \rangle - \langle f \rangle^2)/\langle f^2 \rangle] N_{cc} + 2\Delta f \langle f \rangle \langle f^2 \rangle N_{nc}, \tag{3.27}$$

where the partial coordination numbers N_{ij}, N_i and N_{nc} are defined as:

$$N_{ij} = \int 4\pi r^2 \rho_{ij}(r)\,dr, \tag{3.28}$$
$$N_i = \sum_j N_{ij}, \tag{3.29}$$
$$N_{nc} = \int 4\pi r^2 \rho_{nc}(r)\,dr. \tag{3.30}$$

3.2.5 Topological and Chemical Order in Binary Solutions

A crystalline, disordered alloy, consisting of 1- and 2-type atoms, may be considered as an ideal crystal composed of "average" atoms with superimposed "inhomogeneities" represented by variations in the atomic scattering amplitudes and the static displacements due to the difference in the sizes (size effect) between the real atoms of the solid solution and the average atom on each lattice site. It is often advantageous to consider the diffraction pattern of such disordered crystalline solutions to consist of regular diffraction peaks (Bragg reflections) produced by the average lattice, and a diffuse scattering (Laue diffuse scattering) produced by the fluctuations of the composition and by distortions due to the difference in size of the 1- and 2-type atoms [3.29]. The average lattice describes the positional or topological order which is long-ranged in crystalline solids, and the fluctuations in composition and the distortions are a measure of the chemical order which may be long- or short-range in crystalline solutions.

In amorphous solids, both topological and chemical order extend only over short ranges, and we speak, therefore, of Topological Short-Range Order (TSRO), represented by the number–number correlation function $\rho_{nn}(r)$ and its Fourier transform $S_{nn}(Q)$, and Chemical Short-Range Order (CSRO), represented by the concentration–concentration correlation function $\rho_{cc}(r)$ and its Fourier transform $S_{cc}(Q)$. The number-concentration correlation function $\rho_{nc}(r)$ and its Fourier transform $S_{nc}(Q)$ are related to the size effect [3.26].

The chemical short-rage-order parameter α_1 for the first coordination shell can be determined from the values of N_{ij} and N_i or N_{nn}, N_{cc} and N_{nc}, i.e., [3.31]

$$\alpha_1 = \int \alpha(r)\, dr = \frac{N_{cc}}{(c_2 N_1 + c_1 N_2)} = 1 - \frac{N_{12}}{[c_2(c_2 N_1 + c_1 N_2)]}$$

$$= 1 - \frac{N_{12}}{\{c_2 N_{nn}[1 - (c_1 - c_2)(N_1 - N_2)/N_{nn}]\}}. \quad (3.31)$$

In crystalline solutions, the number N_1 of atoms ("1" or "2") about a "1" atom is the same as the number N_2 of atoms about a "2" atom, i.e. $N_1 = N_2$. This may also be the case in amorphous solids if the component atoms are of similar size. Then, α_1 reduces to the Warren-Cowley short-range order parameter $(\alpha_1)_{wc}$ of the first coordination shell, i.e., [3.29]

$$(\alpha_1)_{wc} = 1 - N_{12}/(c_2 N_{nn}) = N_{cc}/N_{nn}. \quad (3.32)$$

3.3 Diffraction Theory of Powder Pattern Peaks from Nanocrystalline Materials

3.3.1 Fourier Analysis of the Peak Profiles

The total interference function $I(Q)$ of a powder pattern can be expressed as the sum of the interference functions $I_{hkl}(Q)$ of the individual (hkl) reflections multiplied by the Debye-Waller factor $\exp(-\langle u^2\rangle Q^2)$, where $\langle u^2\rangle$ is the mean-square displacements of the atoms from the equilibrium positions due to thermal vibration and static displacements, and the thermal diffuse scattering which modulates about the values of $[1 - \exp(-\langle u^2\rangle Q^2)]$:

$$I(Q) = \sum_{hkl} [I_{hkl}(Q)\exp(-\langle u^2\rangle Q^2)] + [1 - \exp(-\langle u^2\rangle Q^2)]W(Q), \quad (3.33)$$

where $W(Q)$ is a modulation function characteristic of each Bravais lattice [3.29].

It is well known that each profile $I_{hkl}(Q)$ of a powder pattern peak can be expressed as a Fourier series whose coefficients are related to the effective particle size $\langle D_e \rangle$, due to the size of the coherently diffracting domains (grains or subgrains) and stacking faults in these grains, and to microstrains within the small domains in heavily cold-worked materials [3.22, 29, 32].

The function $I_{hkl}(Q)$ can be approximated as

$$I_{hkl}(Q) = I_{measured}(Q)/(\text{peak area}), \quad (3.34)$$

and can be written as Fourier series:

$$I_{hkl}(Q) = \sum_n |C(L)| \exp[iL(Q - Q_0)], \quad (3.35)$$

where $L = nd_{hkl}$ is the distance normal to the reflecting planes (hkl) with the interplanar spacing d_{hkl}, n is the harmonic number of the Fourier series, and Q_0 is the Q-value of the peak maximum position of the (hkl) reflection.

3.3 Diffraction Theory of Powder Pattern Peaks from Nanocrystalline Materials

The Fourier coefficients $|C(L)|$, with $C(0) = 1$, can be expressed as the product of the particle size coefficients $A_p(L)$, due to the effective particle size $\langle D_e \rangle$, and the strain coefficients $|C_s(L)|$ produced by the strain fields of dislocations. For small values of L and $1/d_{hkl}$, one can approximate these coefficients by

$$A_p(L) \approx 1 - L/\langle D_e \rangle \approx \exp(-L/\langle D_e \rangle), \qquad (3.36)$$

$$|C_S(L)| \approx 1 - (2\pi^2 L^2)[\langle(\varepsilon_L)^2\rangle - \langle\varepsilon_L\rangle^2]/(d_{hkl})^2$$

$$\approx \exp\{-(2\pi^2 L^2)[\langle(\varepsilon_L)^2\rangle - \langle\varepsilon_L\rangle^2]/(d_{hkl})^2\}, \qquad (3.37)$$

and

$$\ln|C(L)| = \ln A_p(L) - (2\pi^2 L^2)[\langle(\varepsilon_L)^2\rangle - \langle\varepsilon_L\rangle^2]/(d_{hkl})^2, \qquad (3.38)$$

where the effective particle size $\langle D_e \rangle$ includes the size $\langle D \rangle$ of the coherently diffracting domains, and $[\langle(\varepsilon_L)^2\rangle - \langle\varepsilon_L\rangle^2]$ is a measure of the standard deviation of the strain distribution about the average strain $\langle\varepsilon_L\rangle = (\Delta L/L)_L$, which is small or zero in small powder particles.

Plotting $\ln|C(L)|$ as a function of $1/(d_{hkl})^2$ [$= (h^2 + k^2 + l^2)/a^2$ in cubic materials with lattice parameter 'a'] for constant values of L allows us to determine the particle-size Fourier coefficients $A_p(L)$ and the root mean square strains $[\langle(\varepsilon_L)^2\rangle - \langle\varepsilon_L\rangle^2]^{1/2}$ from the intercept of the straight line with the ordinate and its slope, respectively (Warren-Averbach method) [3.29, 32]. These quantities can be correlated with the dislocation densities produced by the mechanical or cold working of the powders [3.33]. A plot of $\ln|C(L)|$ versus L (3.38) is shown in Fig. 3.14 for ball-milled W powder. The particle-size Fourier coefficients $A_p(L)$ are plotted in Fig. 3.15 [3.34].

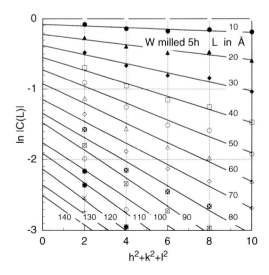

Fig. 3.14. Plot of $\ln|C(L)|$ of W powder, milled for 5 h in a high-energy ball mill, plotted as a function of $h^2 + k^2 + l^2$ for different values of $L = nd_{hkl}$

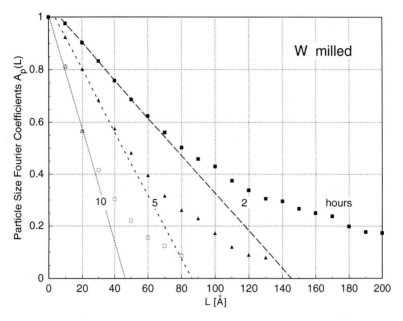

Fig. 3.15. Particle size Fourier coefficients $A_p(L)$ of W milled from 2 to 10 h in a high-energy ball mill, plotted as a function of $L = nd_{nkl}$

3.3.2 Integral Breadth of Powder Pattern Peaks

The integral breadth $b(Q)$ of an (hkl) peak, broadened by small particles and large strains, is defined as the ratio of peak area to maximum peak height. It is related to the Fourier coefficients $C(L)$ by

$$b(Q) = 2\pi / \int C(L)\,dL. \tag{3.39}$$

The integral breadth $b_1(Q)$ due to particles of size D_1 is given $b_p(Q) = 2\pi/D_1$. Thus, the sum of the Fourier coefficients $A_p(L)$ yields the integral breadth particle size D_1, which is related to $\langle D_e \rangle$ by [3.29]

$$D_1 = \int A_p(L)\,dL = \langle (D_e)^2 \rangle / \langle D_e \rangle \geq \langle D_e \rangle. \tag{3.40}$$

If the particle-size coefficients $A_p(L)$ (3.36) can be expressed by the Fourier transform of a Cauchy-Lorentz function and the strain coefficients (3.37) by the Fourier transform of a Gauss function, then the peak profile due to particle size and strains is the convolution of Cauchy-Lorentz and Gauss functions. As demonstrated by *Halder* and *Wagner* [3.35] and described in *Klug* and *Alexander* [3.36], the integral breadth $B(Q)$ can then be related to the particle size breadth $b_p(Q) = 2\pi/D_p$ and the strain breadth [3.32]

$$b_\varepsilon(Q) = 2\varepsilon_p Q = 2(\pi/2)^{1/2}\langle\varepsilon^2\rangle^{1/2} Q \tag{3.41}$$

by a parabolic relationship

$$b_p(Q)/B(Q) = 1 - [b_\varepsilon(Q)/B(Q)]^2. \tag{3.42}$$

It can be expressed in terms D_p and $\langle\varepsilon^2\rangle^{1/2}$, i.e.,

$$2\pi/B(Q) = D_p - 2\pi\langle\varepsilon^2\rangle D_p[Q/B(Q)]^2. \tag{3.43}$$

It should be pointed out that (3.43) applies to the integral breadth $B(Q)$, and not to the half width, i.e., the full width at half maximum height of the peak. When employing (3.43) using the integral breadth $B(Q)$, corrected for instrumental broadening with the relation [3.36, 37]

$$B(Q)/B_{exp}(Q) = 1 - [B_{standard}(Q)/B_{exp}(Q)]^2, \tag{3.44}$$

measures of the particle size D_p and strain $\varepsilon_p = (\pi/2)^{1/2}\langle\varepsilon^2\rangle^{1/2}$ are obtained, which differ from the values of $\langle D_e\rangle$ and $\langle\varepsilon^2\rangle^{1/2}$.

In principle, the values of D_I and D_p should be quite similar if the assumptions leading to (3.43) apply, but different from $\langle D_e\rangle$ as shown in (3.40). If all particles possess the same size, i.e., $D_I/\langle D_e\rangle = \langle(D_e)^2\rangle/\langle D_e\rangle^2 = 1$, the coefficient $A_p(L)$ would follow the relation

$$A_p(L) = 1 - L/\langle D_e\rangle, \tag{3.45}$$

which is the Fourier transform of

$$F(Q) = (1/\langle D_e\rangle)(\sin^2 Q\langle D_e\rangle/2)/(Q/2)^2 \tag{3.46}$$

representing the interference function of a small single crystal with the shape of a cube and edge length $\langle D_e\rangle$ [3.38]. On the other hand, if the peak profiles could be described by a Cauchy-Lorentz function [3.32]

$$F(Q) = 2\langle D_e\rangle/[1 + \langle D_e\rangle^2 Q^2] \tag{3.47}$$

then

$$A_p(L) = \exp(-L/\langle D_e\rangle) \tag{3.48}$$

and $D_I/\langle D_e\rangle = \langle(D_e)^2\rangle/\langle D_e\rangle^2 = 2$.

In general, the profile or a powder pattern peak can rarely be approximated by the convolution of Cauchy-Lorentz and Gauss functions. Thus, it is more advantageous to apply the Fourier-analysis technique to the profile of the powder pattern peaks from cold-worked metals and alloys. It is certainly not more time-consuming than the analysis of the integral breadths. In any event, the use of the full width at half maximum height (commonly called half width) of the powder pattern peak should be avoided, since it will yield unreliable results, e.g., in the case of a bimodal particle-size distribution in the cold-worked samples.

3.4 Experimental Diffraction Techniques

3.4.1 Radiation Sources

X-rays and neutrons are the radiation sources most commonly used in modern scattering experiments. Although electron diffraction experiments were performed in the past to study the structure of amorphous materials, they are rarely employed today.

For many years, X-rays have been generated in sealed and open tubes, using electron beams, accelerated to a maximum of 100 keV and yielding a continuous spectrum and a series of characteristic X-ray wavelengths depending on the anode material. A tremendous increase in X-ray intensity occurred when the synchrotron radiation became available after 1970. This radiation is produced by fast electrons in a synchrotron yielding a continuous spectrum which can be further enhanced by insertion devices [3.39].

Neutron sources have always been mayor installations. With the development of nuclear reactors, neutrons became available in sufficient numbers to permit the performance of neutron scattering experiments [3.40]. More recently, accelerators have also been used to generate neutrons. An intense, pulsed beam of high-energy protons bombards a heavy metal target and produces spallation neutrons. In both reactor and spallation source, the neutrons must be moderated (or slowed down) to velocities or energies to yield wavelengths about 1 Å.

3.4.2 Diffraction Methods

In any diffraction experiment, one tries to measure the intensity scattered by the sample as a function of the length $Q = 4\pi(\sin\theta)/\lambda$ of the diffraction vector **Q** in (3.4). It is readily seen that two experimental conditions can be applied [3.24]:

(1) Variable 2θ method: Q can be varied by changing the diffraction angle 2θ between the incident and scattered beams using monochromatic X-rays or neutrons.

(2) Variable λ method: Q can be changed by using polychromatic X-rays or neutrons and measuring the energy E of the photons or the time-of-flight t of neutrons at a fixed scattering angle 2θ.

3.4.3 Variable 2θ Method

Any modern scattering experiment based on the measurement of the scattered intensity as a function of the diffraction angle 2θ consists of a radiation source, a monochromator to select a narrow band of wavelengths and to suppress unwanted radiation, a diffractometer with a sample holder, and a radiation detector. In the conventional diffractometer technique, a single detector (proportional counter, scintillation counter or solid-state detector) measures the scattered intensity through a narrow collimator sequentially as a function of 2θ, either continuously or in steps, and is controlled by a microprocessor or computer. This technique is still very useful when using X-rays from powerful sources coupled with a sufficient amount of sample material because it provides the best peak-to-background ratio due to the fact that the detector sees only a very small solid angle ($<0.2°$) of the scattered radiation.

A great improvement in sensitivity and counting efficiency was accomplished when it became possible to detect simultaneously all radiation scattered by the sample (similar to the old Debye-Scherrer film technique). The development of linear and curved position-sensitive detectors for X-rays and neutrons [3.41] has advanced to the state that it permits the registration of the diffracted radiation over a large angular range in 2θ, as large as $120°$. It must be emphasized that these detectors should be used with sample and flight paths in an evacuated chamber to reduce parasitic scattering due to the large solid angle of the detector window.

3.4.4 Variable λ Method

With the development of intense sources of continuous radiation for X-rays and neutrons, the measurement of the scattered intensity at a fixed diffraction angle 2θ becomes very attractive. In the case of X-rays, the use of a solid-state detector with an energy resolution of 150 eV at 6 keV makes this technique of energy-dispersive diffraction possible [3.42]. The time structure of the neutrons produced in the pulsed spallation source allows the application of the time-of-flight diffraction [3.43]. In order to make absolute measurements of the scattered intensity, it is necessary to know the wavelength or energy dependence of the primary radiation spectrum. In the case of neutrons, it is possible to use a vanadium sample to determine the wavelength spectrum [3.44]. In the energy-dispersive X-ray measurements, the precise determination of the primary-beam spectrum is rather elaborate. One can use high-angle [3.42] or low-angle scattering [3.45] from an amorphous sample to evaluate the wavelength spectrum of the continuous radiation generated by a conventional X-ray tube.

3.4.5 Analysis of the Diffraction Pattern

a) *Total Diffracted Intensity from Amorphous and Nanocrystalline Samples*

The interference function $I(Q)$ (3.15) or the structure factor $S(Q)$ (3.16) is not a measured quantity, but is deduced from the experimental X-ray intensity after correction for polarization, absorption, and Compton and multiple scattering. These corrections can be carried out by the computer programs, e.g., INTER and FOURIER written in FORTRAN for IBM-compatible computers, based on an earlier program for main frame computers [3.23]. This program not only corrects the raw data, but normalizes them to scattered intensity per atom $I_a(Q)$ [3.23, 24]. Then it calculates both $I(Q)$ and $S(Q)$, and performs the Fourier transforms to obtain the reduced atomic distribution functions $G_l(r)$ (3.17) and $G_S(r)$ (3.18).

b) *Fourier Analysis of the Profiles of Powder Pattern Peaks*

The Fourier analysis of the individual reflections of the nanocrystalline materials is also performed with a personal computer, e.g., using the PASCAL program LBA, based on an earlier FORTRAN program for main frame computers [3.32]. It allows the separation of the K_α profile of each peak into its K_{α_1} and K_{α_2} components, the correction for instrumental broadening using a well-annealed powder as standard, and the separation of the particle size and strains using first- and second-order peak profiles [3.32]. It also calculates the integral breadth and corrects it for instrumental broadening using (3.44). Equation (3.43) is then employed for the separation of particle size and strains.

3.5 Structure of Amorphous and Nanocrystalline Alloys

In order to illustrate the power of the diffraction experiments to elucidate the structure of amorphous alloys prepared by rapid quenching from the liquid state or by mechanical alloying, we will discuss amorphous beryllium-, vanadium- and tungsten-based alloys. It should be remembered that a single diffraction experiment yields only the total structure factors $I(Q)$ (3.15) and $S(Q)$ (3.16) which represent the weighted sums of the three partial structure factors $I_{ij}(Q)$ and $S_{nc}(Q)$, respectively. In order to determine the three individual partial structure factors, three diffraction experiments must be carried out, each with a different set of weight factors $c_i f_i c_j f_j$ (3.12) or $\langle f \rangle^2$ (3.14). This can be accomplished by preparing three different samples by isomorphous or isotopic substitution of one or both alloying elements. However, there are two special cases in neutron experiments, where a single experiment can yield directly one of the three partial functions. If one of the components has a negative scattering length such as Ti, one can prepare a so-called 'zero' alloy where $\langle f \rangle = 0$, e.g., $Be_{36}Ti_{62}$ (Fig. 3.13), which yields directly the concentration–concentration correlation function $\rho_{cc}(r)$, the Fourier transform of the structure factor $S_{cc}(Q)$ (3.7), i.e., the chemical short-range order [3.46]. If the component element 2 has a zero or very small scattering length, such as V for neutrons, any alloy consisting of elements 1 and 2 will yield directly the partial pair distribution function $\rho_{11}(r)$, the Fourier transform of the structure factor $I_{11}(Q)$ (3.13) [3.47]. In the case of X-ray scattering, the structure factor $S(Q)$ of any alloy composed of a light element, such as Li and Be, and a heavy element, such as Zr and W, is dominated by the partial structure factor of the heavy element.

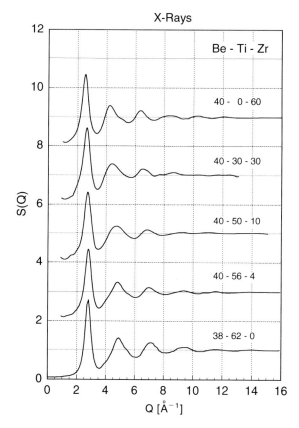

Fig. 3.16. X-ray structure factors $S(Q)$ of Be–(Ti–Zr) alloys [3.27]. The X-ray values of $\langle f \rangle^2/\langle f^2 \rangle$ vary between 0.654 for $Be_{40}Zr_{60}$ and 0.751 for $Be_{38}Ti_{62}$

3.5.1 Amorphous Beryllium Alloys

Amorphous ribbons of Be–Ti–Zr alloys were produced from melts of $Be_{40}Ti_{60-x}Zr_x$ with $x = 4$, 10, 30, 50 and 60 at.% and $Be_{38}Ti_{62}$ by melt-spinning at Lawrence Livermore National Laboratory [3.27]. The X-ray data were obtained with the variable wavelength method (energy-dispersive X-ray diffraction) and the neutron experiments were performed on the Intense Pulsed Neutron Source (IPNS) at the Argonne National Laboratory using the time-of-flight technique.

The total structure factors $S(Q)$ (3.12, 14, 16) are shown in Figs. 3.16 and 3.17. The difference in appearance of $S(Q)$ between the X-ray data and the neutron results is quite striking, and is due to the fact that the weight factors $\langle f \rangle^2/\langle f^2 \rangle$ are quite different for X-rays and neutrons as indicated in Figs. 3.17 and 3.18. Beryllium is practically invisible for X-rays and only the partial functions Zr(Ti)–Zr(Ti) and Zr(Ti)–Be are observable. The neutron scattering lengths

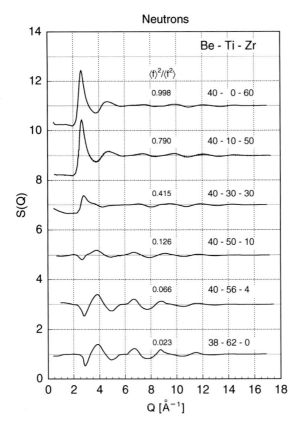

Fig. 3.17. Neutron structure factors $S(Q)$ of Be–(Ti–Zr) alloys [3.27]. Also shown are the values of $\langle f \rangle^2 / \langle f^2 \rangle$

for Be and Zr are positive and similar in magnitude, whereas that of Ti is negative. Thus, $S(Q)$ represents the number-density structure factor $S_{nn}(Q)$ (3.14) and (3.16) for $Be_{40}Zr_{60}$, and $S_{cc}(Q)$ for $Be_{38}Ti_{62}$ [3.46].

The Fourier transforms of $Q[S(Q) - 1]$, i.e., $G_S(r)$ (3.18), of the neutron data are shown in Fig. 3.18. The neutron distribution function $G_S(r)$ for the amorphous $Be_{40}Zr_{60}$ alloy exhibits a first peak with three subsidiary maxima indicating Be–Be, Be–Zr, and Zr–Zr neighbour distances at 2.23, 2.76, and 3.23 Å, respectively, whereas $G_S(r)$ for the amorphous $Be_{38}Ti_{62}$ shows a positive peak at 2.26 Å corresponding to Be–Be nearest neighbours, and a negative peak at 2.60 Å due to Be–Ti neighbours. The X-ray distribution function $G_S(r)$ of the $Be_{38}Ti_{62}$ alloy is dominated by the Ti–Ti pairs and yielded the value of 2.83 Å for the distance between Ti–Ti nearest neighbours [3.27].

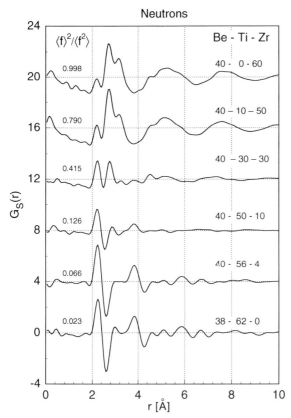

Fig. 3.18. Neutron reduced atomic distribution function $G_S(r)$ of Be–(Ti–Zr) alloys [3.27]. Also shown are the values of $\langle f \rangle^2/\langle f^2 \rangle$

3.5.2 Amorphous and Nanocrstalline Vanadium Alloys

The fact that vanadium has an almost zero scattering length for neutrons has first been utilized by *Lemarchand* et al. [3.48] to determine the partial structure factor $I_{NiNi}(Q)$ in liquid Ni–V alloys. However, no amorphous V alloys have been prepared so far by rapid quenching from the liquid state. Therefore, it was quite a surprise when *Fukunaga* et al. [3.49, 50] reported the amorphization of Cu–V and Ni–V alloys by mechanical alloying, even though the size difference between V and Ni or Cu atoms is only 5%.

The changes in the Cu–Cu partial structure factor $I_{CuCu}(Q)$ and the partial radial distribution function $R_{CuCu}(r) = 4\pi r^2 \rho_{CuCu}(r)/c_{Cu}$ as a function of milling time are shown in Figs. 3.19 and 20, respectively. After 30 h of mechanical alloying, the $Cu_{50}V_{50}$ alloy has a nanocrystalline structure. All *(hkl)* reflections can be recognized in the maxima of $I_{CuCu}(Q)$, and the face-centered coordination shells are visible in the radial distribution function. After 120 h of milling, the

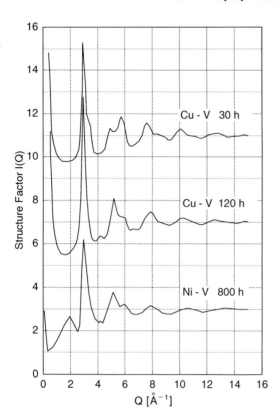

Fig. 3.19. Neutron interference function $I(Q)$ of Cu–V and Ni–V alloys prepared by mechanical alloying [3.49, 50]

diffraction pattern indicates that the structure of the alloy has become amorphous. The corresponding structure factor and the radial distribution function of the $Cu_{50}V_{50}$ sample are shown in Figs. 3.19 and 20, respectively. The disappearance of the 2nd and 5th coordination shells, which are found mainly in the octahedral units of the *fcc* structure, has been used by *Fukunaga* et al. [3.49] to postulate that the amorphous structure consists mainly of tetrahedral units which are responsible for the 3rd, 4th, 6th and 7th coordination shells and are distinguishable in the amorphous structure.

Similar observations were made by *Fukunaga* et al. [3.50] during the amorphization of Ni–V alloys. The $Ni_{40}V_{60}$ alloy slowly changed from a nanocrystalline to an amorphous structure with increasing milling time. The structure factor and the radial distribution function of the Ni–V sample, milled for 800 h, are also shown in Figs. 3.19 and 20, respectively. In addition, they observed that $I_{NiNi}(Q)$ is independent of concentration for the Ni–V alloys with 30 to 55 at.% Ni, an assumption used by *Wagner* and his co-workers [3.23] and *Waseda* [3.51] to evaluate the partial functions in liquid alloys.

3.5 Structure of Amorphous and Nanocrystalline Alloys 79

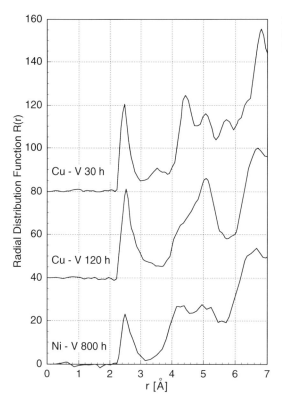

Fig. 3.20. Neutron radial distribution functions of Cu–V and Ni–V alloys [3.49, 50]

3.5.3 Amorphous and Nanocrystalline Tungsten Alloys

Mechanical working or grinding of pure metals and intermetallic phases in a high-energy ball mill has also been employed to produce a nanocrystalline structure [3.52]. Heavy plastic deformation, such as filling, introduces a high density of dislocations in metals and alloys which are responsible for the observed broadening of their powder pattern peaks [3.29, 32]. This broadening has been interpreted in terms of the reduction in size of the coherently diffracting domains and the appearance of microstrains within these domains. There is considerable interest in these materials, since *Gleiter* and his co-workers [3.28, 53] have shown that nanocrystalline metallic powder compacts, produced by evaporation of pure metals in a He atmosphere, collecting the fine particles in a die and compacting them in situ, exhibit interesting physical and mechanical properties.

In order to test the possibility that mechanical working of elemental powders or mechanical alloying of powder mixtures can yield nanocrystalline samples [3.54, 55], *Wagner* et al. [3.34] have chosen W powder to be milled in a high-energy ball mill because W is elastically isotropic and all (*hkl*) reflections

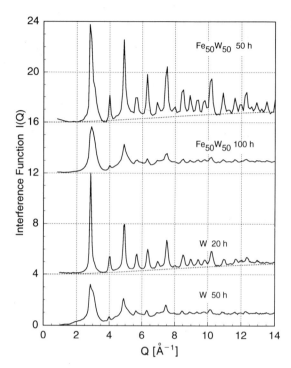

Fig. 3.21. X-ray interference functions $I(Q)$ of W and $Fe_{50}W_{50}$ alloys

can be employed in the Warren-Averbach analysis [3.29, 32] of the diffraction profiles for the separation of particle size and strain broadening (3.38). They found that the particle size decreased to 50 Å after milling for 10 h. However, an amorphous W–Fe phase appeared after 20 h due to the Fe and Cr contamination from the steel vial and stainless steel balls.

In order to study the formation of amorphous W alloys, samples of $Fe_{50}W_{50}$ were prepared by mechanical alloying in the high-energy Spex 8000 mixer/mill. In addition, pure W powder was mechanically worked in the Spex mill. 5 g charges of the powders with particle sizes of 6–9 μm for Fe and 12 μm for W were sealed under argon atmosphere in a hardened steel vial together with six 440C stainless steel balls (4 balls with 6.35 mm diameter and 2 balls with 12.7 mm diameter) and then milled for up to 50 h. The ball-to-powder ratio was thus 4:1. After a milling time of more than 20 h, the particle sizes of the powders were reduced to 1–2 μm, determined from scanning electron micrographs. Energy-dispersive X-ray fluorescence analysis revealed a combined Fe and Cr content of 24 ± 4 at.% after 20 h and 50 ± 10 at.% after 50 h of milling the originally pure W sample. The Fe and Cr uptake was small (<5%) in the $Fe_{50}W_{50}$ alloy even after 100 h of milling.

The diffraction patterns for the pure W powder and the mixture of 50 at.% Fe and W powders showed considerable broadening of the powder pattern peaks even after one hour of milling. After more than 20 h, a broad peak appeared to develop at the position of the (110) reflection of Fe in pure W and the Fe–W

3.5 Structure of Amorphous and Nanocrystalline Alloys

mixture. It corresponds to the first peak in the diffraction pattern from an amorphous Fe–W alloy. Prolonged milling of the powders yielded a two-phase alloy, i.e., a nanocrystalline W solid solution and an amorphous $Fe_{50}W_{50}$ alloy. The peaks in the diffraction patterns correspond to the body-centered cubic Bravais lattice and yield the lattice parameter $a = 3.1525 \pm 0.0005$ Å after 50 h of milling, indicating that a substantial amount of Fe and Cr is dissolved in the W lattice ($a = 3.1653$ Å). Assuming a linear dependence of the atomic volume of a substitutional W–Fe(Cr) alloy with concentration would yield an iron-chromium content of about 6 at.% after 50 h milling.

The total scattering patterns of W(Fe) and $W_{50}Fe_{50}$ were converted into the structure factors or interference functions $I(Q)$ (3.15), which are shown in Fig. 3.21 for milling times of 20, 50 and 100 h. They include the Bragg peaks and the Thermal Diffuse Scattering (TDS) from the nanocrystalline phase and a contribution from the amorphous phase. The intensities of the Bragg reflections are reduced by the Debye-Waller factor $\exp(-\langle u^2 \rangle Q^2)$, where $\langle u^2 \rangle$ is the mean-square displacement of the atoms due to thermal vibration and static displacements. The TDS modulates about the values $1 - \exp(-\langle u^2 \rangle Q^2)$, which are also shown in Fig. 3.21 for the predominantly nanocrystalline samples. A value of $\langle u^2 \rangle = 0.01$ was used in the TDS calculations. It is obvious from Fig. 3.21 that the TDS is largely responsible for the diffuse scattering in the nanocrystalline alloys [3.56, 57].

The Fourier transforms of $I(Q)$, i.e., the reduced atomic distribution functions $G_I(r)$ (3.17) are exhibited in Fig. 3.1. The patterns clearly show that W–Fe possesses the body-centered cubic (bcc) coordination and they are similar in appearance to the face-centered cubic (fcc) powder pattern. Thus, the coordination numbers in the *bcc* lattice are identical to the multiplicity factors of the (*hkl*) reflections of the *fcc* powder pattern.

The size of the nanocrystals can be estimated from the $G_I(r)$ curves. The function $G(r)$ becomes zero as soon as either $\rho(r) - \rho_0$ or $v(r)$ becomes zero beyond a certain value of r (3.17). Since the samples, except those milled for more than 50 h, are crystalline, $v(r)$ must be the cause of the gradual damping of $G(r)$. Taking the value of r beyond which $G(r)$ fluctuates by no more than 5% of the maximum deflection about zero (this is estimated to be caused by the experimental errors in $I(Q)$ due to the statistics of the photon counts and the corrections of the raw data), we deduce the values of the particle or coherently diffracting domain sizes $D_G < 50$ Å, with an estimated error of ± 10 Å, which are shown in Fig. 3.23.

When the peaks of the crystalline phase were removed from the structure factors $I(Q)$ of the alloys milled for 50 and 100 h, depicted in Fig. 3.21, the remaining pattern represented $I(Q)$ of the amorphous W–Fe alloy. The distribution function $G_I(r)$ corresponding to this amorphous W–Fe phase in the samples, which contained about 50 at.% Fe and Cr, is quite similar to that of the entire sample, except that the modulations in $G(r)$ of the amorphous phase do not show much structure beyond 15 Å [3.57]. Since $G_I(r)$ is dominated by the W–W pairs because the weight factors $W_{ij} = c_i f_i c_j f_j / \langle f \rangle^2$ in (3.17) are 0.04, 0.30

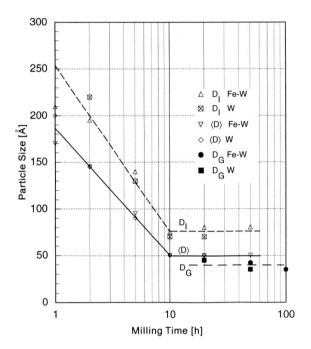

Fig. 3.22. Particle sizes in nanocrystalline W and Fe–W alloys plotted as function of the logarithm milling time in a high-energy ball mill

and 0.66 for Fe–Fe, Fe–W and W–W, respectively, the position of the first peak at 2.70 Å corresponds predominantly to the W–W neighbour distance.

The reflections (110) to (310) of W in the W and $Fe_{50}W_{50}$ samples were subjected to the Warren-Averbach analysis [3.29, 32] (Fig. 3.14). The effective particle size $\langle D_e \rangle$ was determined from the initial slope of the particle size Fourier coefficients $A_p(L)$ (Fig. 3.15) (3.36). The sum of the Fourier coefficients $A_p(L)$ yielded the integral breadth particle size D_I (3.40). The errors in $\langle D_e \rangle$ and D_I are estimated to be ± 5 Å and ± 10 Å, respectively. The values of $\langle D_e \rangle$ and D_I are given in Fig. 3.22 and are similar to the values D_G determined from the atomic distribution function $G_I(r)$. The particle sizes D_I are only slightly larger than $\langle D_e \rangle$ indicating that the particles have a narrow size distribution (3.40).

The microstrains $\langle (\varepsilon_L)^2 \rangle^{1/2}$ at $L = 50$ Å, (3.37), assuming that the average strains $\langle \varepsilon_L \rangle$ are zero or small in the individual powder alloy particles, are shown in Fig. 3.23. Dislocations are introduced by the severe deformation during the mechanical working in the high-energy ball mill and are responsible for the large strains, which increase initially in the W and Fe–W samples, but reach a plateau and then decrease after prolonged milling. The nanocrystalline W particles are embedded in an amorphous W–Fe phase which acts as extended grain boundaries. It is believed that the decrease in strains is caused by the elimination of the excess dislocations in the amorphous phase.

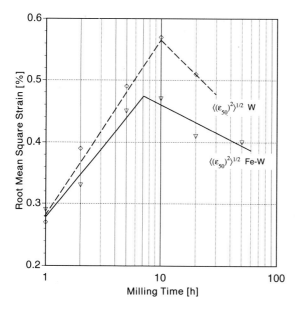

Fig. 3.23. Root-mean-square strains in W and Fe–W alloys plotted as function of the logarithm of the milling time

3.6 Selected Examples of Electron-Microscopy Analysis

Transmission electron microscopy imaging and diffraction can be considered as powerful tools in the investigation of rapidly solidified materials. The extent of these materials includes amorphous, ultrafine grain and rapidly solidified alloys, and Transmission Electron Microscopy (TEM) will provide rapid assessment of a number of these materials' microstructure. The various types of analysis will now be considered briefly.

Perhaps the most easily accomplished electron microscope observation is to use the technique of Selected Area Diffraction (SAD). An appropriate sized aperture is placed in an intermediate imaging plane of the microscope so that the diffraction pattern is produced from only a limited specimen area which can be as small as 2000 Å in diameter. For an amorphous material, a diffraction pattern is produced with broad diffuse rings which is indicative of a material with local short-range order and random longer-range order. An example of a selected area diffraction pattern of an amorphous material consisting of both a copper and carbon mixture is given in Fig. 3.24.

For a polycrystalline specimen, sharp diffraction rings are produced similar to an X-ray powder pattern. An example of a polycrystalline pattern of copper is shown in Fig. 3.25. Accurate measurements of the ring diameters yield the interplanar spacings of the material which can be correlated to tabulated

Fig. 3.24. SAD pattern of an amorphous carbon–copper thin film

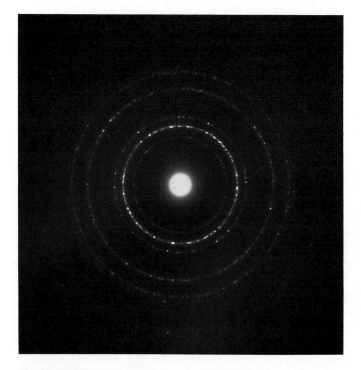

Fig. 3.25. SAD pattern of a polycrystalline Cu thin film

diffraction data (JCPDS tables). These tables are generally for X-ray scattering and the listed relative intensities are in the regime of single scattering events (kinematical scattering approximation). For electrons, however, the scattering power is much larger than for X-rays ($\approx 10^4$) and multiple scattering occurs in all but the thinnest specimens. It is for this reason that a critical evaluation of electron diffraction intensities requires a dynamical diffraction theory treatment and an accurate knowledge of the specimen thickness. Furthermore, inelastic scattering phenomena can dominate the diffraction intensity producing such phenomena as Kikuchi lines. This subject requires specialized theoretical treatments and will not be pursued further here.

In addition to amorphous and polycrystalline materials, there remains the possibility of diffraction from extended single crystals. In this case, well-defined diffraction orders are produced which can be indexed and related to lattice spacings and crystal symmetries characteristic of the material. An example of a quasicrystalline SAD pattern of an Al–Cu–Fe alloy with sharp Bragg type peaks is shown in Fig. 3.26. Here five-fold symmetry is revealed by the pattern which is characteristic of this new class of materials.

Fig. 3.26. Diffraction pattern of a quasicrystal showing strong Bragg-type diffraction orders. The *labeled* reflections from the center correspond to *d*-spacings of 5.3, 3.4, 2.1 and 1.3 Å

86   3 Structure and Characterization of Rapidly Solidified Alloys

It must be pointed out that electron diffraction generally does not have the accuracy found with X-rays in determining d-spacings to 4 or 5 significant figures. The main reason for this is the fact that electrons scatter mainly in the forward direction by only a few degrees at most where very small angular measurements are inherently inaccurate. This can be seen from the Bragg relationship,

$$n\lambda = 2d \sin \theta,$$

where electrons have much shorter wavelengths (λ) than X-rays on the order of 100-fold. However, on a positive note, electron diffraction is performed very rapidly and can be implemented on very small specimens which is not possible with X-rays. Also, coupled with imaging, electron diffraction provides a close look a local diffraction anomalies which would be averaged by the larger statistical sampling using X-ray methods.

Fig. 3.27. High-resolution micrograph of a grain boundary in gold. Here, the image has been noise filtered and contrast enhanced. The three sets of lattice spacings on either side of the interface are 1.43 Å. The projected atomic positions of the gold lattice are superimposed on the microscope image

Electron microscope imaging is also important for rapidly solidified materials. There are different levels of resolution which enter into the considerations of the type of analysis which can be undertaken which is, in turn, dependent on the type of phase of the material considered (amorphous, polycrystalline or single crystalline). Generally, the newer types of electron microscopes which operate at intermediate accelerating voltages between 200 and 400 keV provide high resolution where lattice spacings near the 2 Å level can be directly visualized. This means that there is the possibility of visualizing the atomic arrangements in crystalline and polycrystalline materials. This is particularly valuable when looking at the lattice mismatches of interfacial structure and defects such as, dislocations, inclusions and vacancies as well as stacking faults and twins. An example of an interfacial boundary in a metal is given in Fig. 3.27, where the projected atomic arrangement of the interface is shown superimposed. Generally, this type of structural analysis is dependent on a full understanding of the phase-contrast imaging process in the microscope. Here, the direct transmitted beam and one or more diffracted beams are combined in the image by adjustment of the objective lens' defocus. At the correct value of defocus, this

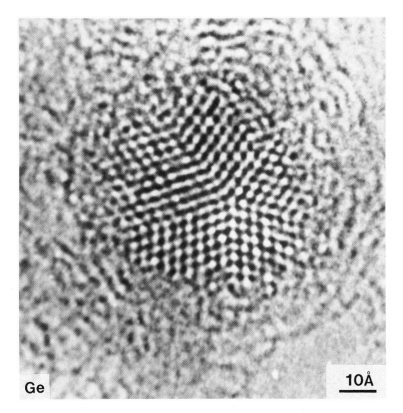

Fig. 3.28. Micrograph of a germanium cluster exhibiting five-fold symmetry

produces an interference pattern which represents the object's structure and is often referred to as a "structure image". An example of visualization of a small germanium cluster is shown in Fig. 3.28, where dark-image features represent atom pairs separated by 1.35 Å. Here, the ≈ 3.2 Å lattice planes of germanium are readily observed and the multiple twinning produces an apparent five-fold symmetry cluster.

In the case of analyzing images it is often necessary to perform dynamical diffraction and image computations to confirm and match the observed structures. It is possible to achieve this, provided the projection of the structure has a relatively simple repeat along the direction of the electron beam. An example of such a computer image simulation is shown in Fig. 3.29b for an Al–Cu–Fe quasicrystal and its projected atomic structure is shown in Fig. 3.29a.

Although it has not been mentioned to this point, it is possible to image defects in crystals by diffraction contrast instead of lattice imaging. In this case a particular Bragg reflection is selected and included in the microscope objective aperture to produce either a bright-field or dark-field images. The strain field of the crystalline materials **R** is very sensitive to the particular Bragg reflection represented by the vector **g**. The image amplitude of a defect is proportional to the exponential of $(2\pi i/\lambda)(\mathbf{g} \cdot \mathbf{R})$. It is possible to obtain large contrast effects by this method. However, the technique is limited in resolution to perhaps 20 Å by the objective aperture size and Fresnel diffraction in the crystal. Displacements of a fraction of a lattice spacing are easily detected by this method which is far more sensitive than direct lattice imaging. An example of this type of image contrast is given in Fig. 3.30, where the strain field at atomic steps in the boundary produces a large contrast effect of three lobes at each step.

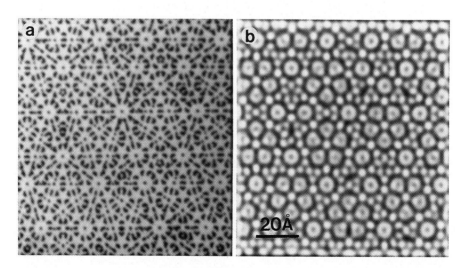

Fig. 3.29. Model (**a**) and computed image (**b**) of a quasicrystalline structure viewed along the five-fold axis

3.6 Selected Examples of Electron-Microscopy Analysis   89

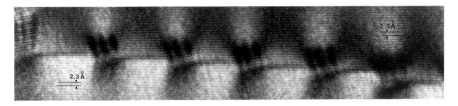

Fig. 3.30. Image of a grain boundary exhibiting strong diffraction contrast (*dark lobes*) at single atomic steps

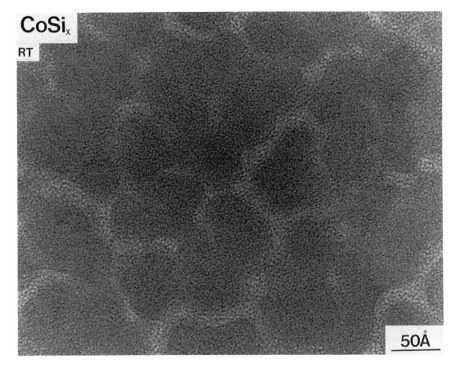

Fig. 3.31. Image of an amorphous thin film of cobalt silicide exhibiting phase contrast granularity. The extended contrast variations are due to changes in silicon concentration

Finally, imaging the amorphous state which often occurs in rapidly solidified materials must be mentioned. Generally, at low resolutions little information is obtained since most materials of this type are relatively structureless until resolutions of nanometer dimensions are approached. Bright-field phase-contrast imaging (the direct beam and the diffuse scattering) usually yields an image which appears to be composed of a random granular pattern of black-and-white image features of various sizes down to the resolution of the microscope. It is well known that this "phase contrast granularity" represents how information is transmitted by the electron microscope, i.e., its transfer function. If the microscope is properly adjusted for optimum information transfer an image results which represents the projected structure. Unfortunately, for amorphous materials atoms do not align in columns along the beam direction, as in crystals, and hence, the resultant random pattern occurs. It is useful to inspect this type of image to confirm if a material is truly amorphous or whether microcrystallization has occurred. An example of an amorphous thin film is displayed in Fig. 3.31.

Here, a very brief explanation of some aspects, of electron microscopy imaging and diffraction only suffices to show its potential as an analytical tool. It remains for the practiced investigator to make this tool part of the arsenal used to characterize materials of interest.

For further information about electron microscopy and high-resolution imaging techniques please see [3.58–69].

References

3.1 T. Egami, R.S. Williams: IEEE Trans. MAG-**12**; 927 (1976)
3.2 T. Egami, K. Maeda, V. Vitek: J. Philos. Mag. A **41**, 883 (1980)
3.3 K. Maeda, S. Takeuchi: Phys. Status Solidi (a) **49**, 685 (1978)
3.4 K. Maeda, S. Takeuchi: J. Physique **12**, 283 (1978)
3.5 S.A. Rice, P. Gray: *The Statistical Mechanics of Simple Liquids* (Wiley, New York 1965)
3.6 T. Egami, D. Scrolovitz: J. Phys. F **12**, 2163 (1982)
3.7 D. Srolovitz, K Maeda, V. Vitek, T. Egami: Philos. Mag. A **44**, 847 (1981)
3.8 M.A. Otooni: Processing path and the evolution of crystallinity. *Rapidly Solidified Glassy Alloys*. MRS Proc. **362**, 101–109 (1995)
   M.A. Otooni: A phenomenological approach to the ductility in ultra-fine grain. *Rapidly Solidified Materials*. MRS Proc. **362**, 49–59 (1995)
   M.A. Otooni: Explosive consolidation of pulverized rapidly solidified materials, in Adv. Mater. and Proc. PRICM-1, Huangzhou, PR. China (1992) Tech. Dig. pp. 123–130
   M.A. Otooni: Critical radii and the onset of supercoling in rapidly solidified alloy systems, Jpn. MRS Conf. on Adv. Mater. (Tokyo 1989)
   M.A. Otooni: Kinetics of crystallization of the amorphous Cu_{60}-Zr_{40} alloy system, ARDEC Tech. Rept. SCS-R-SR-8-810 (1984)
   M.A. Otooni: J. Non-Cryst. Sci. **65**, 389-402 (1984)
3.9 M.A. Otooni: J. Non-Cryst. Solids **61 & 62**, 1347–1352 (1984)
3.10 M.A. Otooni: Laser annealing and induced short range ordering in the Cu_{60}-Zr_{40} alloy system. J. Mater. Res. Soc. **23**, 345–352 (1984)
3.11 M.A. Otooni: Thermal cycling and the onset of the amorphous-crystalline transition in the Cu_{60}-Zr_{40} alloy system. MRS Proc. **8**, 277–282 (1982)
3.12 E. Blanzat: Mater. Sci. Eng. **23** (10), 151 (1976)
3.13 F.E. Luborsky, J.J. Beker, R.O. McCray: IEEE Trans. MAG-**13**, 1644 (1975)
3.14 H.H. Liebermann, C.D. Graham, P.J. Flanders: IEEE Trans. MAG-**13**, 1541–1547 (1977)
3.15 S. Aur, T. Egami, I. Vincze: Amorphous metals, in *Proc. Int'l Conf. on Rapidly Quenched Metals*, Sendai, Jpn. (1978) Paper RQ-4, p. 1081
3.16 E. Svab: J. Non-Cryst. Solids **46**, 276 (1981)
3.17 G.C. Chi, H.S. Chen, C.E. Miller: J. Appl. Phys. **49**, 946–962 (1978)
3.18 N.A. Pratten, M.G. Scott; Script. Met. **12**, 137 (1978)

3.19 B. Cantor, F. Duflos: Martensite in splat-quenched iron and iron nickel, in *Rapidly Quenched Metals III*, ed. by B. Cantor (Metals Soc., Metals City, OH 1978) pp. 110–118
3.20 B. Cantor, Y. Inokuti: Script. Met. **10**, 655 (1976)
3.21 R.W. Cahn, M. Krishnan, M. Laridjani, J. Greenholz, R. Hill: *Proc. 2nd Int'l Conf. on Rapidly Quenched Metals*, ed. by N.J. Grant, B.C. Giessen (Elsevier, Lausanne 1976) Vol. 2, p. 83
3.22 C.N.J. Wagner: In *Microscopic Methods in Metals*, ed. by U. Gonser, Topics Curr. Phys., Vol. 40 (Springer, Berlin, Heidelberg 1986)
3.23 C.N.J. Wagner: In *Liquid Metals. Chemistry and Physics*, ed. by S.Z. Beer (Dekker, New York 1972) p. 257
3.24 C.N.J. Wagner: J. Non-Cryst. Solids **31**, 1 (1978)
3.25 A.B. Bathia, D.E. Thornton: Phys. Rev. B **2**, 3004 (1970)
3.26 A.C. Wright, C.N.J. Wagner: J. Non-Cryst. Solids **106**, 85 (1988)
3.27 S.B. Jost, A.E. Lee, C.N.J. Wagner, L.E. Tanner: Z. Phys. Chem. (Neue Folge) **157**, 11 (1989)
3.28 X. Zhu, R. Birringer, U. Herr, H. Gleiter: Phys. Rev. B **35**, 908 (1987)
3.29 B.E. Warren: *X-Ray Diffraction* (Addison-Wesley, New York 1968)
3.30 E.A. Lorch: J. Phys. C **2**, 229 (1969)
3.31 C.N.J. Wagner: In *Proc. 5th Int'l Conf. Rapidly Quenched Metals*, ed. by S. Steeb, H. Warlimont (North-Holland, Amsterdam 1984) p. 405
3.32 C.N.J. Wagner: In *Local Atomic Arrangements Studied by X-Ray Diffraction*, ed. by J.B. Cohen, J.E. Hillard (Gordon & Breach, New York 1966) p. 219
3.33 R.P.I. Adler, H.M. Otte, C.N.J. Wagner: Met. Trans. **1**, 2791 (1970)
3.34 C.N.J. Wagner, E. Yang, M.S. Boldrick: Adv. X-Ray Anal. **35**, 585 (1992)
3.35 N.C. Halder, C.N.J. Wagner: Adv. X-Ray Anal. **9**, 91 (1966)
3.36 H.P. Klug, L. Alexander: *X-Ray Diffraction Procedures for Polycrystalline ans Amorphous Materials*, 2nd edn. (Wiley, New York 1974) p. 661
3.37 C.N.J. Wagner, E.N. Aqua: Adv. X-Ray Anal. **7**, 46 (1964)
3.38 A. Guinier: *Principles of X-Ray Diffraction* (Freeman, New York 1963)
3.39 J.E. Spencer, H. Winick: In *Synchrotron Radiation Research*, ed. by H. Winick, S. Doniach (Plenum, New York 1980) p. 662
3.40 R. Pynn: Rev. Sci. Instrum. **55**, 837 (1984)
3.41 R.C. Hamlin (ed.): Trans. Am. Cryst. Assoc., Vol. 18 (1982)
3.42 T. Egami: In *Glassy Metals*, ed. by H.J. Güntherodt, H. Beck, Topics Appl. Phys., Vol. 46 (Springer, Berlin, Heidelberg 1981) Chap. 3
3.43 J.D. Jorgenson, A.J. Schultz (eds.): Trans. Am. Cryst. Assoc. **29** (1993)
3.44 R.N. Sinclair, D.A.G. Johnson, J.C. Dore, H.H. Clarke, A.C. Wright: Nucl. Instrum. Methods **117**, 445 (1974)
3.45 G. Fritsch, C.N.J. Wagner: Z. Physik B **62**, 189 (1986)
3.46 C.N.J. Wagner: J. Non-Cryst. Solids **150**, 1 (1992)
3.47 C.N.J. Wagner, M.S. Boldrick: J. Alloys Comp. **194**, 295 (1993)
3.48 J.L. Lemarchand, J. Bletry, P. Desre: J. Physique **41**, C8-163 (1980)
3.49 T. Fukunaga, M. Mori, K. Inou, U. Mizutani: Mater. Sci. Eng. A **134**, 863 (1991)
3.50 T. Fukunaga, V. Homma, K. Suzuki, M. Misawa: Mater. Sci. Eng. A **134**, 987 (1991)
3.51 Y. Waseda: *The Structure of Non-Crystalline Materials* (McGraw-Hill, New York 1980)
F.E. Fujita: *Physics of New Materials*, Springer Ser. Mater. Sci., Vol. 27 (Springer, Berlin, Heidelberg 1994)
3.52 R.B. Schwarz, C.C. Koch Appl. Phys. Lett. **49**, 146 (1986)
3.53 R. Birringer, H. Gleiter: In *Encyclopedia of Materials Science*, ed. by R.W. Cahn (Pergamon, Oxford 1988) Suppl. Vol. l, p. 339
3.54 E. Hellstern, H. Fecht, Z. Fu, W.L. Johnson: J. Appl. Phys. **65**, 305 (1988)
3.55 H. Fecht, E. Hellstern, Z. Fu, W.L. Johnson: Met. Trans. **21**, 2333 (1990)
3.56 C.N.J. Wagner, M.S. Boldrick: Mater. Sci. Eng. A **133**, 26 (1991)
3.57 C.N.J. Wagner, E. Yang, M.S. Boldrick: Nano-Structured Mater. **7**, 1 (1966)
3.58 A.D. Beukel, J. Sietsma: Acta Met. Mater. **38**, 383–389 (1990)
3.59 E.N. Sheftel, D.E. Kaputkin, R.E. Stroug: MRS Proc. **362**, 137–142 (1995)

3.60 H. Jones: Growth parameters in formation and stability of rapidly solidified microstructure, in *Science and Technology of Rapid Solidification and Processing*, ed. by M.A. Otooni (Kluwer, Dordrecht 1995) pp. 13–24
3.61 G.S. Shiflet, Y. He, S.J. Poon, G.M. Dougherty, H. Chen: In *Science and Technology of Rapid Solidification Technology and Processing*, ed. by M.A. Otooni (Kluwer, Dordrecht 1994) pp. 53–73
3.62 W. Krakow, D.A. Smith J. Mater. Res. **1**, 47 (1986)
3.63 W. Krakow, J.T. Wetzel, D.A. Smith: Philos. Mag. A **53**, 739 (1986)
3.64 W. Krakow, D.P. DiVincenso, P.A. Bancel, E. Cockayne, V. Elser: J. Mater. Res. **8**, 24 (1993)
3.65 W. Krakow: Ultramicrosc. **45**, 269 (1992)
3.66 W. Krakow Philos. Mag. A **63**, 233 (1991)
3.67 W. Krakow: J. Electron Microsc. Tech. **17**, 212 (1991)
3.68 W. Krakow MRS Proc. **237**, 447 (1992)
3.69 L. Reimer: *Transmission Electron Microscopy*, 4th edn., Springer Ser. Opt. Sci., Vol. 36 (Springer, Berlin, Heidelberg 1997)

4 Atomic Transport and Relaxation in Rapidly Solidified Alloys

H. Kronmüller

Diffusion processes in solids play a central role in tailoring high-quality materials for technical applications. Whereas in crystalline solids a number of intrinsic self-diffusion processes could be identified previously, it is just recently that the situation in amorphous materials could be clarified. In crystalline solids self-diffusion may be controlled by vacancies in thermal equilibrium (noble metals, transition and refractory metals) [4.1, 2], by self-interstitial atoms as in Si [4.3], or by direct atomic exchange and by ring mechanisms as discussed recently for alkaline metals and some transition metals [4.2, 4]. None of these mechanisms seems to apply directly to amorphous alloys. Diffusion in amorphous alloys has been reviewed by several authors concentrating on different aspects. *Cahn* [4.5], *Egami* [4.6], and *Egami* and *Waseda* [4.7] considered the relations between diffusional properties and the glass-forming ability. *Johnson* [4.8] considered the relations between diffusion and diffusion-controlled amorphization processes. *Cantor* [4.9] emphasized that the diffusivity of metals and metalloids in glasses scales with the glass transition temperature. *Taub* and *Spaepen* [4.10] studied the relations between viscosity and diffusional properties, *Kronmüller* and co-workers [4.11–13] as well as *Faupel* et al. [4.14] concentrated on the self-diffusion of metallic components and magnetic and mechanical relaxation phenomena.

Diffusion in amorphous alloys has been discussed sometimes controversially because the number of precise experimental results was restricted. Since amorphous alloys correspond to metastable structures of a certain composition of alloying components these materials tend to transform and separate into more stable phases. The underlying diffusion processes naturally correspond to irreversible phenomena because the structural changes cannot be removed by a cyclic process. Nevertheless, under certain conditions the self-diffusion coefficient in amorphous alloys may be treated as a reversible material property depending on the thermodynamic variables of state, as temperature T, and pressure p. This behaviour requires a stabilization of the amorphous materials at high temperatures, however, below the crystallization temperature. In such pretreated materials diffusion coefficients or relaxation times are reversible material properties being fully determined by a thermodynamic equation of state. In unstabilized materials, however, diffusion coefficients and relaxation phenomena have a strong irreversible component because of quenched-in free volumes (quasivacancies) which enhance the diffusion process. If the diffusing species disappear partially or if they change their microstructure this, in general, corresponds to an irreversible structural change. This behaviour is related to the metastable structure of amorphous alloys. Therefore, diffusion data of amorphous alloys oftenly are discussed controversial depending on the pre-treatment and also on the measuring technique applied. Whereas crystalline materials are characterized by unique diffusion parameters, e.g., unique activation enthalpies, amorphous materials have to be described by effective diffusion parameters depending on temperature and the time interval for which they are determined. An alternative point of view of this behaviour is that of short-range and long-range diffusion processes. Therefore, measurements at low temperatures or short times lead to different results than measurements at high temperatures or long times because in these cases different parts of the activation enthalpy spectrum have become active and contributed to the measured change of a physical property. The following topics will be discussed:

(*i*) Irreversible and reversible diffusional properties;
(*ii*) Self-diffusion measurements by the radiotracer technique;
(*iii*) Analysis of the self-diffusion mechanisms in amorphous alloys;
(*iv*) Effective-medium theory of self-diffusion in disordered media;
(*v*) Magnetic and elastic relaxation effects.

4.1 Basic Equations of Diffusion

Diffusion in solids usually is described by combining phenomenological equations with atomic considerations of the transport of matter. The phenomenological description starts by introducing the atomic concentration, c, of the diffusing species. Under the influence of thermal fluctuations, the mobile species of an atomic structure may change their position. In thermal equilibrium and a spatially homogeneous distribution of the concentration c, the average displacement R of a mobile species after some time t from its original position according to Einstein's law is given by

$$\langle R^2 \rangle = 2dDt, \tag{4.1}$$

where d denotes the dimensionality of the problem and D the, so-called, diffusion coefficient. If the original distribution of the concentration, $c(\mathbf{r})$, contains gradients there exists a current $\mathbf{j}$ of the mobile species into the direction of the negative particle gradient which is described by Fick's first law

$$\mathbf{j} = -D \, \mathrm{grad} \, c(\mathbf{r}). \tag{4.2}$$

If, in addition, the mobile species are exposed to a field of force derived from an interaction potential $E(\mathbf{r})$, Fick's first law reads

$$\mathbf{j} = -D \left(\mathrm{grad} \, c + \frac{c}{kT} \mathrm{grad} \, E \right). \tag{4.3}$$

The so-called second Fick's law is obtained from the continuity equation

$$\frac{\partial c}{\partial t} + \text{div}\,\mathbf{j} = 0 \tag{4.4}$$

by inserting (4.2) or (4.3). In the general case where D is independ on space and the interaction E obeys the Laplace equation $\Delta E = 0$ we obtain from (4.4)

$$\frac{\partial c}{\partial t} = D\left(\Delta c + \frac{1}{kT}\nabla c \cdot \nabla E\right). \tag{4.5}$$

In the case of a one-dimensional problem without any field of force, (5) reads

$$\frac{\partial c(x,t)}{\partial t} = D\frac{\partial^2 c(x,t)}{\partial x^2}. \tag{4.6}$$

A special solution of (4.6) is the so-called thin-film solution. If a nanolayer of marked atoms (radio-tracer atoms) diffuses from a planar surface into a specimen we obtain the distribution $c(x)$ from the solution [4.15, 16]

$$c(x,t) = \frac{c^*}{\sqrt{\pi Dt}}\exp\left(-\frac{x^2}{x_D^2}\right) = \frac{c^*}{\sqrt{\pi Dt}}\exp\left(-\frac{x^2}{4Dt}\right), \tag{4.7}$$

where c^* denotes the area density of the marked atoms, t the diffusion time and x the distance from the surface. The distribution $c(x)$ corresponds to a Gaussian and the average penetration depth, x_D, after time t corresponds to the two-dimensional Einstein equation (4.1)

$$x_D = 2\sqrt{Dt}. \tag{4.8}$$

If during the diffusion process the diffusion coefficient changes, e.g., by a change of the diffusing species it is useful to consider the time average [4.17, 18]

$$\langle D(t,T)\rangle = \frac{1}{t}\int_0^t D_{\text{mom}}(t,T)\,dt' \tag{4.9}$$

of the momentary diffusion coefficient, $D_{\text{mom}}(t,T)$. Equation (4.6) now reads

$$\frac{\partial c(x,t)}{\partial(\langle D\rangle t)} = \frac{\partial^2 c(x,t)}{\partial x^2} \tag{4.10}$$

with the thin-film solution

$$c(x,t) = \frac{c^*}{\sqrt{\pi\langle D\rangle t}}\exp\left(-\frac{x^2}{4\langle D\rangle t}\right). \tag{4.11}$$

Equation (4.11) shows that also in the case of a time-dependent diffusion coefficient a Gaussian distribution exists. From the time-dependent average diffusion coefficient $\langle D\rangle$, the momentary diffusion coefficient may be determined by

$$D_{\text{mom}}(t,T) = \frac{d}{dt}(\langle D(T,t)\rangle t). \tag{4.12}$$

Equation (4.11) has to be applied if reversible or irreversible changes of the density or the microstructure of the diffusing species take place during the diffusion experiment. In the case of solid materials, crystalline or amorphous ones, in general, the diffusion coefficients of diffusing species can be described by Arrhenius equations

$$D(T) = D_{00} \exp[(-H_{\text{eff}} + TS_{\text{eff}})/kT] = D_0 \exp(-H_{\text{eff}}/kT), \qquad (4.13)$$

where $G_{\text{eff}} = H_{\text{eff}} - TS_{\text{eff}}$ denotes an effective free activation enthalpy and H_{eff} the activation enthalpy. The pre-exponential factor

$$D_0 = D_{00} \exp(S_{\text{eff}}/k) \qquad (4.14)$$

contains the activation entropy, S_{eff}, with

$$D_{00} = g a^2 v_0, \qquad (4.15)$$

and a the lattice constant, v_0 the attempt frequency of the order of the Debye frequency, $g = z/6$ for cubic lattices with z the number of nearest neighbours. In the case of crystalline materials where self-diffusion oftenly is governed by a unique mechanism, H_{eff} and S_{eff} correspond to well-defined properties of the periodic lattice potential. It has been a big surprise that within certain temperature ranges also in amorphous solids, where the diffusing species are submitted to a statistical potential, a unique effective activation enthalpy is observed. In the general case, H_{eff} is expected to be temperature dependent, and then $H_{\text{eff}}(T)$ is defined as

$$H_{\text{eff}}(T,t) = -\left(\frac{\partial}{\partial(1/kT)}\right) \ln D(T,t). \qquad (4.16)$$

The validity of Arrhenius laws for the diffusion coefficients is related to the jump frequencies v_{ij} of the diffusing species between an initial position i and a neighbouring position j. If h_i denotes the enthalpy at site i and h_{ij} the enthalpy of the saddlepoint between i and j, v_{ij} is given by

$$v_{ij} = v_{ij,0} \exp[-(h_{ij} - h_i)/kT], \qquad (4.17)$$

where $v_{ij,0}$ contains the migration entropy and the Debye frequency (4.14) and (4.15). It is the task of diffusion theory to determine the relation between D_0, H_{eff} and the statistical quantities h_i, h_{ij} and $v_{ij,0}$.

4.2 Self-Diffusion in Amorphous Alloys

4.2.1 Radiotracer Technique

Amorphous ribbons were produced by the melt-spin technique (width 10 mm, thickness 20–40 µm). The surface of the circular discs prepared from these

4.2 Self-Diffusion in Amorphous Alloys   97

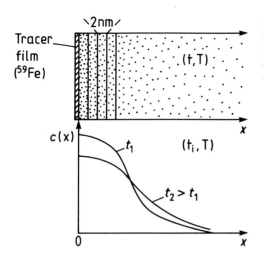

Fig. 4.1. Formation of a radiotracer profile from a monolayer by an annealing treatment. The diffusion profiles (*lower*) are determined by intersectionning of the specimen into 2 nm layers

ribbons was carefully smoothened by lapping and polishing. By chemical deposition $a \approx 1$ nm film of the radiotracer element (e.g., ^{59}Fe, ^{95}Zr) was deposited on the smoothened side. A diffusion profile of the radiotracer element, as shown schematically in Fig. 4.1, was produced by annealing the specimen in an evacuated quartz ampoule. The diffusion profiles were determined by microsectioning the specimen by ion-beam sputtering and subsequent measuring the radioactivity of the sputtered-off section which has been collected on a polymere ribbon (Fig. 4.1 and 2, and Appendix). By this method the achieved precision of the sectioning was 1 nm. This allows the determination of the diffusion coefficient D with an accuracy of half-an-order-of-magnitude [4.17–20].

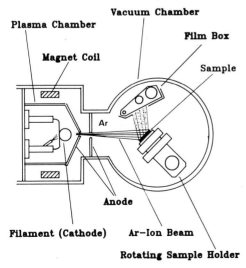

Fig. 4.2. Schematic presentation of the sputtering procedure by Ar ions. Each intersectional film is stored on the polymere ribbon

Self-diffusion experiments in amorphous alloys are performed in the temperature range from 500–800 K in time intervals from 300 to 10^5 s. The diffusion profiles extend over a depth of 10–1000 nm.

4.2.2 Non-Equilibrium and Quasi-Equilibrium of Diffusional Properties

In as-quenched melt-spun amorphous alloys irreversible structural changes take place at elevated temperatures resulting in considerable changes of magnetic, mechanical and electrical properties. Consequently, in this state the self-diffusion coefficient D cannot be considered as an intrinsic reversible thermodynamic property. The observed changes of two orders of magnitude in D may be attributed either to a change of the density of quasi-vacancies assisting the diffusion process or to a change of the activation enthalpy due to irreversible changes of the structure. Accordingly, D is not only a function of the thermodynamic variables T and p, but also of the pre-treatment. In the as-quenched state free volumes or quasi-vacancies are present in supersaturation which above Room Temperature (RT) enable a rearrangement of the atomic structure until a quasi-equilibrium, also denoted as an isoconfigurational state [4.10], is established. These excess quasi-vacancies lead to an enhancement of the diffusion coefficient as in crystalline materials if vacancies are produced by fast particle irradiation. In this case, the diffusion coefficient according to

$$D(T,t) = D_R(T) + \delta D(T,t) \tag{4.18}$$

is composed of the relaxed part $D_R(T)$ and a part $\delta D(T,t)$ representing the enhancement of the diffusion coefficient due to the excess free volumes. In Fig. 4.3, the effect of an annealing treatment at 360 °C on the diffusion profiles of the radiotracer ^{59}Fe in Fe$_{91}$Zr$_9$ is displayed for different annealing times. Following (4.7) from a logarithmic plot

$$\ln c(x) \quad \text{versus} \quad x^2/t, \tag{4.19}$$

the quantity $1/4\langle D \rangle$ may be determined from the straight lines if a Gaussian profile develops. According to Fig. 4.3, the slopes of the obtained straight lines increase with increasing annealing time until after an annealing time of ≈ 26 h the diffusion coefficient approaches a constant value because the structure has lost the excess free volumes.

The existence of a metastable thermodynamic equilibrium of the amorphous structure, where the physical properties become reversible thermodynamic properties, is demonstrated in Fig. 4.4 where the changes of D during a cyclic temperature-time process are shown [4.21]. Figure 4.4 proves that the saturation value (equilibrium value) of D depends only on T and not of the diffusion time applied. After an 18 h annealing at 633 K (state 1) the Fe$_{91}$Zr$_9$ specimen was coated with a radioactive ^{59}Fe monolayer and then diffusion annealed at 673 K (state 2 in Fig. 4.4). The state 2 also was achieved after a diffusion treatment of a non-preannealed specimen (state 3 in Fig. 4.4).

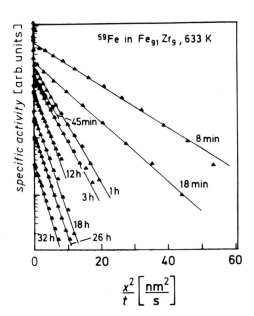

Fig. 4.3. Diffusion profiles of ^{59}Fe radiotracers at 360 °C in amorphous $Fe_{91}Zr_9$ (x = penetration depth, t = annealing time for the development of the profile [4.18, 19])

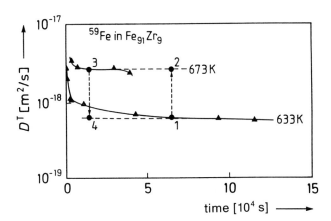

Fig. 4.4. Prove for the existence of quasi-equilibria in the pre-annealed state by cyclic measurements of the radiotracer diffusion

Annealing the specimen of state 3 at 633 K for 1 h led to a state 4 which corresponds to the saturation value of state 1 as to be expected for a reversible thermodynamic material property. This rather time-consuming experiment clearly proves the existence of reversible material properties if the amorphous alloys are stabilized by an annealing treatment at higher temperatures than the measuring temperature.

4.2.3 Review of Diffusion Data

It has to be considered as a great surprise that the radiotracer self-diffusion coefficient in relaxed amorphous alloys obeys an Arrhenius equation [4.5, 9, 11–14, 20, 21]

$$D_R^T(T) = D_{R,0}^T \exp(-H_R/kT). \tag{4.20}$$

Figure 4.5 summarizes a large number of Arrhenius plots, $\log D$ vs $1/T$, of radiotracer diffusion coefficients of a series of amorphous alloys. The diffusion of Fe and Zr has been investigated in Fe-rich as well as Zr-rich FeZr alloys. Large variations in the activation enthalpies and the pre-exponential factors can be observed, e.g., $H_R = 1.5\,\text{eV}$, $D_{R,0}^T = 10^{-7}\,\text{m}^2/\text{s}$ for Fe diffusion in $Fe_{91}Zr_9$, and $H_R = 3.2\,\text{eV}$, $D_{R,0}^T = 10^7\,\text{m}^2/\text{s}$ for Zr diffusion in $Zr_{76}Fe_{24}$. From these results it is obvious that the activation parameters depend on the size of the diffusing atoms as well as on the matrix within which these atoms are diffusing. Figure 4.5 shows for $Fe_{91}Zr_9$ that an Arrhenius equation holds over a temperature range of 300 °C. Furthermore, it is of interest that the diffusion coefficient in crystalline α-Fe is a factor of 10^8 smaller than the Fe diffusion in $Fe_{91}Zr_9$ [4.11].

By a study of the self-diffusion of the three transition metals Fe, Co and Ni in the amorphous alloy $Co_{58}Ni_{10}Fe_5Si_{11}B_{16}$ it could be shown that only small variations of H_R and $D_{R,0}$ take place because the atomic radii (1.25–1.28 Å) of these elements are rather similar [4.22]. H_R varies from 2.11 to 2.24 eV and $D_{R,0}$ changes within the range $10^{-5}\,\text{m}^2/\text{s} < D_{R,0} < 10^{-3}\,\text{m}^2/\text{s}$. In contrast, the diffusion coefficient of Ta is two orders of magnitude smaller with the larger diffusion

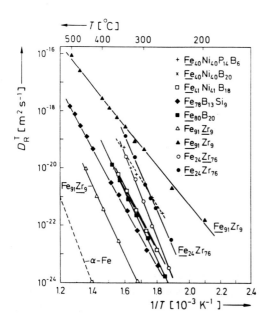

Fig. 4.5. Arrhenius plots of radiotracer diffusion coefficients, D_R^T, in relaxed amorphous alloys. The diffusing components are *underlined* [4.11, 12]; (---) self-diffusion coefficient of α-Fe

4.2 Self-Diffusion in Amorphous Alloys   101

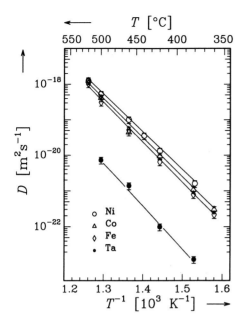

Fig. 4.6. Arrhenius plots of the radiotracer self-diffusion coefficients of Fe, Co, Ni and Ta in relaxed $Co_{58}Ni_{10}Fe_5Si_{11}B_{16}$ (○ Ni, △ Co, ◇ Fe, ● Ta) [4.22]

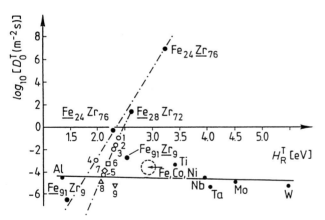

Fig. 4.7. Empirical relationship (4.21) between the diffusion enthalpies and the pre-exponential factors, $D_{R,0}^T$, of transition metals in amorphous alloys and of refractory metals and Al. [● FeZr alloys, (1) $Fe_{80}B_{20}$, (2) $Fe_{40}Ni_{40}B_{20}$, (3) $Fe_{41}Ni_{41}B_{180}$, (4) $Fe_{40}Ni_{40}P_{14}B_6$, (5) $Fe_{78}Bi_{13}Si_9$, (6) Ni, (7) Co, (8) Fe, (9) Ta in $Co_{58}Ni_{10}Fe_5Si_{11}B_{16}$]

enthalpy of $H_{\text{eff}} = 2.37$ eV and a smaller $D_{R,0}$ of 5×10^{-6} m²/s (Fig. 4.6). This difference with respect to Fe, Co and Ni has to be attributed to the larger atomic radius of 1.43 Å of Ta.

In Fig. 4.7 for several amorphous alloys the pre-exponential factors $D_{R,0}^T$ are plotted vs the activation enthalpies H_R. As a general trend an increase of $D_{R,0}^T$ with increasing H_R is observed. For the amorphous FeZr alloys an empirical relationship [4.11, 12]

$$\ln D_R^T(T) = A(H_R - B) \tag{4.21}$$

holds, with $A = 8$ (eV)$^{-1}$ and $B = 2.4$ eV. The pre-exponential factor varies for the amorphous alloys from 10^{-7} to 10^7 m^2/s, whereas the pre-exponential factors of crystalline metals – low melting (Al) and refractory metals (W) – according to Fig. 4.7 vary only in the range between 10^{-5} and 10^{-4} m^2/s. The slight decrease of $D_{R,0}^T$ with increasing activation enthalpy in this case may be attributed to the decrease of the Debye frequency with increasing atomic order number. It is furthermore of interest to note that the diffusion enthalpies of transition metals in a transition-metal-rich amorphous matrix are 0.5–1.0 eV smaller than in crystalline transition metals.

4.2.4 Diffusion Mechanisms in Amorphous Alloys

The self-diffusion coefficients of amorphous alloys show all features of a thermally activated process and in addition in the relaxed state obey reversible thermodynamics (Fig. 4.4). It is therefore self-suggesting to consider similar diffusion mechanisms as in crystalline metals where the dominating diffusion process has been found to be an *indirect diffusion mechanism*, due to vacancies in thermal equilibrium. As shown in Fig 4.8, in this case the vacancy plays the role of a vehicle which enables the hopping of one of the neighbouring atoms into the site of the vacancy which now occupies a position of one of the original nearest neighbour atoms. Figure 4.8 presents also the *direct diffusion mechanism* due to interstitial atoms. In this case, no assisting vehicle is required. A direct diffusion mechanism is also the *direct exchange* or a *ring mechanism* between neighbouring atoms.

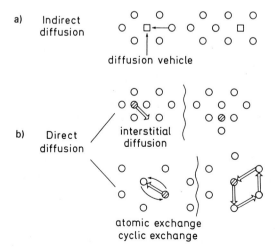

Fig. 4.8. a Indirect diffusion mechanism. **b** Direct diffusion mechanisms; interstitial, exchange and ring mechanisms

In the case of unrelaxed alloys, the enhanced diffusivity is attributed to excess free volumes. The contribution δD to the diffusion coefficient introduced in (4.18) therefore is related to an indirect diffusion mechanism. This also is supported by the fact that the annealing of δD takes place by an activation enthalpy of the order of 1.2 eV. In contrast to crystalline metals, however, the excess quasi-vacancies responsible for δD anneal during the relaxation process (Fig. 4.4) until a quasi-equilibrium of the amorphous structure is reached. The diffusion mechanism in the relaxed material then exhibits many features of a direct diffusion mechanism:

1) The diffusion enthalpy is about 0.5–1.0 eV smaller than in crystalline metals. This fact suggests that the diffusion enthalpy is equal to the migration enthalpy only and not to the sum of formation and migration energy as in crystalline metals.

2) A very instructive experiment concerning the diffusion mechanism is the investigation of the irradiation-enhanced diffusion. From crystalline materials it is well known [4.23] that the increase of the density of vacancies by irradiation leads to a decrease of the effective activation enthalpy. In the case where we deal with an indirect diffusion mechanism where diffusion vehicles are required, e.g., self-diffusion via vacancies in thermal equilibrium, the radiation-induced vacancies would reduce the effective activation enthalpy, H_{eff}, because these excess vacancies do not require the energy consumption of the formation enthalpy. On the other hand, no decrease of H_{eff} is expected if we deal with a direct diffusion mechanism because in this case no diffusion vehicles have to be produced by thermal activation. However, the pre-exponential factor is supposed to be increased due to the larger number of diffusing defect configurations. Both effects mentioned so far have been observed [4.24–26].

A decreasing activation enthalpy under Kr^+ irradiation has been observed in amorphous $Ni_{50}Zr_{50}$ [4.24] and in amorphous FeNiB under Ni^+ irradiation [4.25]. Here, it should be noted that a decrease of H_{eff} is not a necessary prove for an indirect diffusion mechanism in the relaxed material because heavy-particle irradiation may introduce new species of diffusing particles far from any thermodynamic equilibrium.

A constancy of the activation enthalpy and an increase of $D_{R,0}^T$ under proton irradiation of $Co_{58}Ni_{10}Fe_5Si_{11}B_{16}$ has been observed recently [4.26]. Figure 4.9 shows a parallel shift of the Arrhenius plot measured under a proton irradiation with a flux of $\phi = 9 \times 10^{16}\,\text{m}^{-2}\text{s}^{-1}$. The radiation damage produced by 400 keV protons used in this experiment consists mainly in close quasi-Frenkel pairs composed of a quasi-vacancy and a quasi-interstitial atom. At the temperature of irradiation of 300–460 °C, these quasi-Frenkel pairs have a finite life time because their configurations are smeared out. Thus, during proton irradiation a steady-state concentration of smeared out quasi-vacancies of the Frenkel pairs is established. This also is supported by the increase of the diffusion coefficient with increasing radiation flux $\dot{\phi}$. At a diffusion temperature of 420 K according to Fig. 4.10 the diffusion coefficient increases by a factor of 10 if $\dot{\phi}$ increases by a factor of 10 [4.26]. Also, measurements at 380 °C reveal a similar

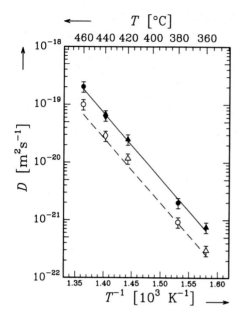

Fig. 4.9. Arrhenius plots of the radiotracer ^{59}Fe self-diffusion coefficient in $Co_{58}Ni_{10}Fe_5Si_{11}B_{16}$ in the relaxed state (*open symbols*) and during irradiation by a proton flux of $\dot{\Phi} = 9 \times 10^{16}\,m^{-2}\,s^{-1}$ [4.26]

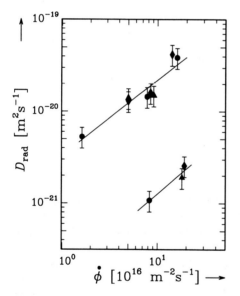

Fig. 4.10. Enhancement of the radiotracer self-diffusion coefficient by a proton flux as measured at 380 °C (*lower points*) and 420 °C (*upper points*) [4.26]

behaviour. These results are compatible with a direct diffusion mechanism. In particular, the increase of D with flux $\dot{\phi}$ has to be attributed to the increase of the steady-state concentration of smeared out quasi-vacancies.

3) Further experiments supporting a direct diffusion mechanism are measurements of the diffusion coefficient during a sudden change of the temperature. Here, it turns out that sudden changes of temperature also lead to a sudden change of the diffusion coefficient [4.27, 28]. In the case of an indirect diffusion mechanism, D_R should change retarded because the formation of a new equilibrium of diffusing vehicles requires relaxation times comparable with the annealing times of δD which would be easily detectable ($t_a \approx 3$ h).

4) A further support for a direct diffusion mechanism in relaxed alloys comes from measurements of the activation volume of the diffusion process. From fundamental thermodynamics it may be derived that the activation volume is approximately given by

$$V = +\left(\frac{\partial G}{\partial P}\right)_P \cong -kT\left(\frac{\partial \ln D}{\partial P}\right)_T, \qquad (4.22)$$

where $G = H - TS$ denotes the Gibbs free energy of activation, and a minor contribution, $kT(\partial \ln D_{00}/\partial P)_T$ has been omitted [4.29]. Similar to the Gibbs free activation energy of an indirect diffusion mechanism, the activation volume according to

$$V = V_f + V_m \qquad (4.23)$$

is composed of contributions due to the formation and the migration of the defects. In the case of a vacancy, V_f is of the order of an atomic volume, Ω_0, whereas V_m is suggested to be as small as 10 to 15% of Ω_0 [4.30]. According to these ideas, the activation volume of a direct diffusion mechanism should be much smaller than Ω_0. Measurements of the diffusion coefficients as a function of pressure have been performed for the Co diffusion in $Co_{76.7}Fe_2Nb_{14.3}B_7$ [4.31, 32]. From the linear relation between $\ln D$ and $1/T$, $H_{eff} = 2.3$ eV and $D_0 = 10^{-3}$ m^2/s have been obtained, and from the pressure dependence of D according to (4.22) an activation volume of $V = -(0.06 \pm 0.1)\Omega_0$ has been derived. A vanishing activation volume also has been obtained for $Fe_{39}Ni_{40}B_{21}$ [4.33]. In contrast, in the case of fcc crystalline Co, the activation volume is found to be $V = 0.71\Omega_0$ [4.34]. Similarly in silver, gold and lead activation volumes of 0.72, 0.72 and 0.71 Ω_0 have been determined [4.35–37]. The measured activation volume of $-0.06\Omega_0$ of amorphous CoFeNbB clearly indicates that the diffusion mechanism is not related to defects in thermal equilibrium which act as vehicles of the diffusion process.

5) Measurements of the isotope effect, i.e., the dependence of D on the atomic mass allow to distinguish between indirect and direct diffusion mechanisms in amorphous alloys. In the case of an indirect diffusion mechanism, an individual atom jumps into a vacancy. The pre-exponential factor, D_{00}, then is proportional to the Debye frequency, i.e., $D_{00} \propto 1/\sqrt{M}$, where M denotes the atomic mass. On the other hand, a direct diffusion mechanism in a nearly close-packed amorphous alloy is not determined by a single atomic jump, but

rather by a cooperative motion of a cluster of n atoms. In this latter case, the dependence of D_{00} on the atomic mass is reduced by a factor of n. Measurements of the isotope effect have been performed for the diffusion of the tracer atoms ^{57}Co and ^{60}Co in relaxed amorphous $Co_{76.7}Fe_2Nb_{14.3}B_7$ [4.31, 38, 39]. Only a minor isotope effect could be detected which was a factor of 7 smaller than that measured for crystalline fcc Co. Accordingly, also the lacking isotope effect supports the direct diffusion mechanism in amorphous alloys. Here it should be noted that in unrelaxed metallic glasses an isotope effect has been detected [4.39], which supports the quasi-vacancy mechanism governing the inversible changes of the diffusion coefficients [4.21].

4.3 Theory of Diffusion in Disordered Media

4.3.1 The Effective-Medium Approximation

The treatment of diffusion of atoms in a statistical random potential is based on the following assumptions: (i) The jump frequency of an atom from a site i to a neighbouring site j obeys an Arrhenius law

$$v_{ij} = v_{ij,0} \exp[-(h_{ij} - h_i)/kT], \tag{4.24}$$

where the attempt frequency $v_{ij,0}$ contains the migration-entropy term, h_i denotes the enthalpy at site i, and h_{ij} the enthalpy at the saddle point between i and j, respectively. (ii) The statistical potential is assumed to be invariant with respect to motions of the mobile species. (iii) The property to be determined is the diffusion coefficient, which, in general, depends on the time t, and is defined as

$$D(t) = \frac{1}{2d} \frac{\partial}{\partial t} \langle R^2 \rangle = \frac{1}{2d} \frac{\partial}{\partial t} \sum_i \mathbf{R}_i^2 P_i(t). \tag{4.25}$$

Here, $\mathbf{R}_i$ are the site vectors and $P_i(t)$ denotes the conditional probability of occupation of site i at time t after starting at $t = 0$ at the origin $\mathbf{R}_i = 0$. The motion of the mobile species is governed by a Markoffian master equation for the occupation probability

$$\frac{d}{dt} P_i(\mathbf{r}_i, t) = \sum_j^{n.n.} [v_{ji} P_j(\mathbf{r}_j, t) - v_{ij} P_i(\mathbf{r}_i, t)], \tag{4.26}$$

where the summation extends over the nearest neighbours (n.n.) of the sites i and j.

4.3.2 Explicite Solutions

Explicite solutions of (4.26) are well-known for periodic crystals where the atomic jump frequency is characterized by a unique activation enthalpy h_{ij} –

$h_i = \Delta H$ and a unique pre-exponential factor $v_{0,ij} = v_0$, which leads to a time-independent diffusion coefficient

$$D = \frac{z}{2d} a^2 v_0 \exp(-\Delta H/kT), \qquad (4.27)$$

(a: distance between, and z: number of nearest neighbours). The situation is much more complex in the case of amorphous alloys. Exact solutions are known for the so-called Random-Trap Model (RTM) [4.40, 41] for $d = 1, 2, 3$ and for the Random-Barrier Model (RBM) for $d = 1$ [4.42, 43]. Explicite results also can be derived if the fluctuations $\delta h_{ij} = h_{ij} - H_0$ around the average activation enthalpy, H_0, obey a Gaussian distribution

$$p(\delta H_{ij}) = \frac{1}{\sqrt{2\pi H_m^2}} \exp(-\delta h_{ij}^2/2H_m^2), \qquad (4.28)$$

where H_m^2 corresponds to the variance of h_{ij}. Using (4.25) different results for D are obtained for the short- and the long-time approximation [4.11]:

(i) Short-time approximation $\langle v_{ij} \rangle t < 1$

$$D^S(T) = a^2 \langle v_{ij,0} \rangle \exp(-H_0/kT) \exp[H_m^2/2(kT)^2]; \qquad (4.29)$$

(ii) Long-time approximation $\langle 1/v_{ij} \rangle^{-1} t \gg 1$

$$D^L(T) = a^2 \frac{1}{\langle 1/v_{ij,0} \rangle} \exp(-H_0/kT) \exp[-H_m^2/2(kT)^2]. \qquad (4.30)$$

Equation (4.29) and (4.30) describe the diffusion coefficients for two limiting cases. At small times only the high frequencies, i.e., the low activation enthalpies contribute to D, whereas at large times all high frequency jumps have died out and the jumps with the large activation enthalpies govern the diffusion. Therefore, different types of averaging have to be applied. Due to the temperature dependence of the effective activation enthalpies which, according to (4.16), are given by

$$H_{\text{eff}}^{S,L} = H_0 \mp \frac{H_m^2}{kT} \qquad (4.31)$$

the Arrhenius plots become nonlinear, where the curvatures increase with increasing variance H_m^2.

4.3.3 The Effective-Medium Approximation for Direct Diffusion Mechanisms

A very effective method to determine diffusion coefficients of disordered systems is the so-called Effective-Medium Approximation (EMA) as developed by several workers. This method, in particular, allows a treatment of the random-trap–random-barrier model [4.43] (Fig. 4.11). In general, all jump frequencies depend on the nearest-neighbour (n.n.) sites i and j and on the direction of the transition. Introducing Laplace transforms

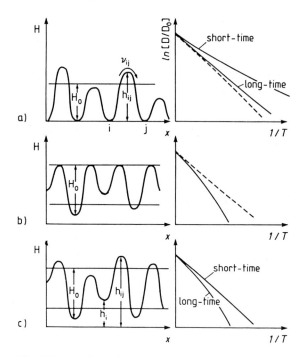

Fig. 4.11a–c. Schematic representation of different random diffusion potentials and the corresponding Arrhenius plots of self-diffusion coefficients; (a) random barrier model, (b) random trap model, (c) mixed model. The *horizontal lines* in the H–x-diagrams correspond to $H = 0$, $H = \langle h_i \rangle$ and $H = \langle h_{ij} \rangle$, respectively (see text)

$$\tilde{P}_i(\mathbf{r}_i, s) = \int_0^\infty P_i(\mathbf{r}_i, t) e^{-st} \, dt, \tag{4.32}$$

we obtain from (4.26) a set of coupled linear equations

$$s\tilde{P}_i(\mathbf{r}_i, s) - \delta_{i,0} = \sum_{j \neq i} [v_{ji} \tilde{P}_j(\mathbf{r}_j, s) - v_{ij} \tilde{P}_i(\mathbf{r}_i, s)], \tag{4.33}$$

where $\delta_{i,0} = 1$ for $i = 0$ and $\delta_{i,0} = 0$ for $i \neq 0$. Equation (4.33) may also be written for the configurational average $\tilde{G}_i(\mathbf{r}_i, s) = \langle \tilde{P}_i(\mathbf{r}_i, s) \rangle_{\{v_{ij}\}}$ for all configurations $\{v_{ij}\}$. By introducing an effective frequency-dependent and space-independent jump frequency, $v_c(s)$, we obtain

$$s\tilde{G}_i(\mathbf{r}_i, s) - \delta_{i,0} = \tilde{v}_c(s) \sum_{j \neq i} [\tilde{G}_j(\mathbf{r}_j, s) - \tilde{G}_i(\mathbf{r}_i, s)]. \tag{4.34}$$

The solution of (4.34) may be obtained by means of Fourrier transforms

$$\tilde{G}_i^*(\mathbf{k}, s) = \sum_i \tilde{G}_i(\mathbf{r}_i, s) \exp(i\mathbf{k} \cdot \mathbf{r}_i), \tag{4.35}$$

which leads to

$$\tilde{G}_i^*(\mathbf{k}, s) = \left(s + \tilde{v}_c(s) \sum_{j \neq i} \{1 - \exp[-i\mathbf{k}(\mathbf{r}_j - \mathbf{r}_i)]\}\right)^{-1}. \tag{4.36}$$

The effective frequency has to be determined from the self-consistency condition that the configurational averages $\langle \tilde{P}_i(\mathbf{r}_i, s) \rangle_{\{v_{ij}\}}$ and $\tilde{G}(\mathbf{r}_i, s, \tilde{v}_c)$ are equivalent, i.e.,

$$\langle \tilde{P}_i(\mathbf{r}_i, s) \rangle_{\{v_{ij}\}} = \tilde{G}_i(\mathbf{r}_i, s, \tilde{v}_c). \tag{4.37}$$

Naturally, it is not possible to express the average $\langle P_i \rangle$ for a real system explicitly as a function of the v_{ij} and the Green's function $\tilde{G}_i$. Therefore, as shown in Fig. 4.12, a cluster of bonds with transition rates corresponding to the original configuration at a position i' is considered. This cluster is embedded within the effective medium described by the effective frequency $\tilde{v}_c$. Equation (4.26) then reads

$$\frac{d}{dt} P_{i'}(\mathbf{r}_i, t) = \sum_{j}^{n.n.i'} [v_{ji'} P_j(\mathbf{r}_j, t) - v_{i'j} P_{i'}(\mathbf{r}_{i'}, t)]$$
$$+ \tilde{v}_c(t) \sum_{i \neq j \neq n.n.i'} [G_j(\mathbf{r}_j, t) - G_i(\mathbf{r}_{i'}, t)]. \tag{4.38}$$

A solution of the self-consistency equation (4.37) allows a calculation of the time-dependent diffusion coefficient, $D(t)$. The Laplace transform of $D(t)$ can be directly determined from

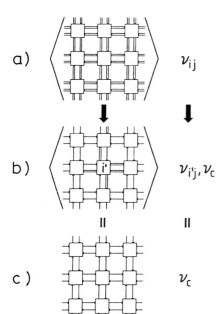

Fig. 4.12a–c. The three different regions of the effective medium approximation; (a) real medium with random jump frequencies, (b) cluster of random v_{ij}'s at i' and effective jump frequency v_c at all other sites; (c) effective medium with jump frequency v_c

$$\tilde{D}(s) = \frac{s}{2d}\langle \tilde{r}_{(s)}^2\rangle = -\frac{s}{2d}\nabla_k^2 \tilde{G}^*(\mathbf{k},s)|_{k=0} = \frac{z}{2d}a^2 \tilde{v}_c(s)/s. \tag{4.39}$$

In the limit of short times ($s > \infty$), this gives

$$D(0) = \lim_{t\to\infty} D(t) = \frac{z}{2d}a^2 \tilde{v}_c(\infty), \tag{4.40}$$

and for the long-time limit, we obtain

$$D(\infty) = \lim_{t\to\infty} D(t) = \frac{z}{2d}a^2 \tilde{v}_c(0). \tag{4.41}$$

4.3.4 Applications of the "Effective-Medium Approximation"

The EMA has been applied to the three types of diffusion potentials shown in Figs. 4.11a–c [4.12]. (1) The Random-Barrier Model (RBM) with h_i independent of i and a Gaussian distribution of h_{ij} with variance h_{RB}^2. (2) The Random-Trap Model (RTM) with h_{ij} independent of i and j and a Gaussian distribution $p(h_i)$ with variance h_{RT}^2. (3) The Mixed Model (MM) with independent Gaussian distributions $p(h_i)$ and $p(h_{ij})$ and variances $h_{RB}^2 = h_{RT}^2$. The temperature dependences of the effective diffusion coefficients of these three models have been determined for the limiting cases of short times (st) and long times (lt) and for a topological structure of a three-dimensional fcc-like disordered structure. The averaging indicated in (4.37) has been reduced to a cluster of a single-bond–single-site model where only one equilibrium site and one saddle point has been varied according to Gaussian distributions. In Fig. 4.11, the broken lines represent the Arrhenius plots for the average activation enthalpy $H_0 = \langle h_{ij} - h_i\rangle$. In the RBM model, the Arrhenius plots for the st and lt case lie above the Arrhenius plot for H_0 since the three-dimensional diffusion path avoids the higher barriers. This effect is more dominant in the st case than the lt case. The effective diffusion enthalpy in the st case is given by $H_{\text{eff}} = H_0 - h_{RB}^2/kT$. In the RTM model, $D(t)$ is time independent and lies below the average Arrhenius plot because the diffusing specy cannot avoid the low-lying traps in the case where v_{ij} is independent on h_j. The lt-effective diffusion enthalpy is given by $H_{\text{eff}} = H_0 + h_{RT}^2/kT$. In the MM model, the st limit coincides with the Arrhenius plot for H_0 because the fluctuations of h_{ij} and of h_i just cancel each other for $h_{RB}^2 = h_{RT}^2$. In the case of the lt limit, $D(t)$ lies below the H_0 Arrhenius plot because of the dominance of the deep traps at long times. It is evident that the three-dimensional MM model is most suitable to describe diffusion in a three-dimensional random potential as it is expected to describe an amorphous alloy. Therefore, time-averaged Arrhenius plots have been determined for different measuring times which are shown in Fig. 4.13. The st line is identical with the Arrhenius plot for an average activation enthalpy $H_0 = 1.65\,\text{eV}$ and $v_{0,ij} = 10^{12}\,\text{s}^{-1}$. For the calculations of Arrhenius plots, for the measuring times $t =$

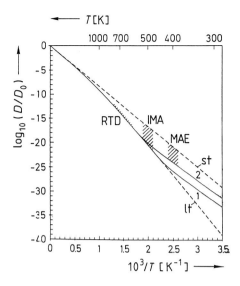

Fig. 4.13. Arrhenius plots of time averaged diffusion coefficients for the mixed model and for the long-time (lt) and the short-time (st) limit. *Curve 1* for $t = 10^5$ s and *curve 2* for $t = 10^2$ s. The *hatched regions* indicate the regimes where RTD, *induced magnetic anisotropies* MA and MAE experiments have been performed [4.11]

10^2 s (curve 2) and $t = 10^5$ s (curve 1) and for the lt limit, the variances $h_{RB} = h_{RT} = 0.22$ eV have been used. Figure 4.13 is suitable for the interpretation of different types of diffusion experiments made in different temperature ranges and time intervals. The hatched regions in Fig. 4.13 indicate the temperature and time intervals where radiotracer, Induced Magnetic Anisotropy (IMA) and Magnetic After-Effect (MAE) experiments usually are performed. In agreement with Fig. 4.14, where we represent the activation enthalpies obtained with these three techniques, the largest values for H_{eff} are that of the RTD followed by the spectra of activation enthalpies obtained for the IMA and the MAE. Here, it

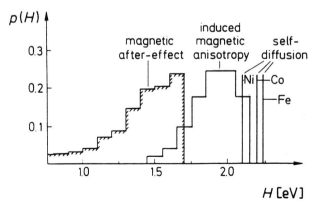

Fig. 4.14. Self-diffusion enthalpies of Fe, Co, Ni in $Co_{58}Ni_{10}Fe_5Si_{11}B_{16}$ as obtained from RTD [4.21], and spectra of activation enthalpies obtained from IMA and MAE. The average measuring temperatures and times were 700 K/10^5 S (RTD), 500 K/10^5 s (IMA) and 400 K/10^2 s (MAE)

should be noted that the results obtained by the EMA are comparable to the results obtained by Monte-Carlo simulations [4.44].

The very surprising fact that in the case of RTD long-time experiments, the diffusion coefficients are described by unique activation enthalpies (Fig. 4.5) may be understood within the framework of the EMA. In the case of the RBM and the single-site–single-barrier approximation, the self-diffusion coefficient for the lt limit is given by

$$D(T) = \frac{z}{2d} a^2 v_c(T), \qquad (4.42)$$

where v_c for $z \geq 2$ obeys the self-consistency condition

$$\left\langle \frac{v_{i',j} - v_c}{v_{i',j} + \left(\frac{z}{2} - 1\right) v_c} \right\rangle_{jn.n.i'} = 0. \qquad (4.43)$$

Introducing the distribution function $\rho(v_{i'j})$ (normalized to one atom) for the transition frequencies and $v_{i'j} = v_0 \exp(-h_{i'j}/kT)$, (4.43) reads

$$\frac{2}{z} = \int_0^\infty dh_{i'j} \rho(h_{i'j}) \{1 + \exp[(h_{i'j} - \mu)/kT]\}^{-1}. \qquad (4.44)$$

The expression in the square brackets of (4.44) corresponds to the Fermi function with a Fermi energy

$$\mu = -kT \ln[(z/2 - 1) v_c/v_0], \qquad (4.45)$$

where μ has the meaning of an effective barrier height, defining

$$v_c(T) = \frac{v_0}{z/2 - 1} \exp(-\mu/kT). \qquad (4.46)$$

Since the Fermi function for $T \to 0$ can be replaced by a step function, μ is determined by

$$\frac{2}{z} = \int_0^\mu \rho(h_{i'j}) \, dh_{i'j}. \qquad (4.47)$$

4.3.5 Molecular Dynamics Simulations and Diffusion Mechanisms

The atomistic processes underlying the phenomenological EMA theory has been studied also by Molecular Dynamics Simulations (MDS) [4.45, 46]. These calculations are based on electron-tight-binding atom pair potentials [4.47] which are adjusted so as to reproduce the mass density, the compressibility and the glass temperature of the amorphous alloy. The motion of the individual atoms is determined using the extended-system-method of Nosé [4.48] which allows an isobaric-isothermal treatment of the system. The investigations have been performed for the amorphous FeZr alloys. In the iron-rich alloy $Fe_{91}Zr_9$, a small value of $D_0 = 10^{-7} m^2/s$ is observed in the case of the Fe diffusion, whereas in $Fe_{24}Zr_{76}$ large values of $D_0 = 10^7 m^2/s$ for the diffusion of Zr are measured. These large values of D_0 for the Zr diffusion are due to the requirement of a collective motion of a large number of neighbouring atoms which also explains the large activation enthalpy of $H \cong 3.2$ eV as well as a large positive migration entropy. The situation is different in the case of Fe diffusion in the iron-rich matrix $Fe_{91}Zr_9$. Because the iron atoms have a smaller ionic radius than the Zr atoms they may jump into neighbouring quasi-vacancies with a relatively small activation enthalpy of 1.5 eV. The small D_0 values in this case are attributed to a negative migration entropy, because the quasi-vacancy on the n.n. site of the hopping atom has to be produced by an agglomeration of spread-out free volumes by statistical thermal fluctuations as shown in Fig. 4.15. This process is, however, not equivalent to the formation of vacancies in crystalline lattices where an indirect diffusion mechanism by means of vacancies in thermal equilibrium is valid. In the case of amorphous alloys, free volumes are present as a prerequisite of the amorphous structure, and their agglomeration does not require a formation energy. The MDS is suitable to test whether the motion of atoms in amorphous alloys may be characterized by atomic jumps as in crystalline materials. The development of the displacement of an atom has been determined for a melt, a metglass and a crystalline structure. Whereas in the melt, the displacement takes place continuously with time, in the metglass and the crystal the displacement takes place discontinuously by distances of 2–3 Å. In the crystalline structure, one or two oscillation periods are required for a jump, and in the amorphous structure 5–10 local oscillations, because a number of metastable intermediate configurations have to be overcome. A detailed study shows that the postulated collective motion takes place by thermally activated displacement chains. Such a displacement chain where six atoms are involved is shown in Fig. 4.16. These jumps with displacements of ≈ 1.5 Å take place within a time interval of 800 fs at temperatures of 1382 °C. In Fig. 4.17, the final positions of each of the six atoms are marked. Atom 1 occupies the position of atom 2 which then occupies the position of atom 3, etc. The last atom of N atoms involved finally occupies a position which has been

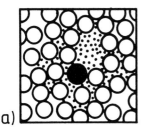 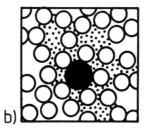

Fig. 4.15. a Direct diffusion of small Fe atoms into a quasi-vacancy produced by agglomeration of free volumes. **b** Collective motion of large Zr atoms by means of smeared-out free volumes

114   4 Atomic Transport and Relaxation in Rapidly Solidified Alloys

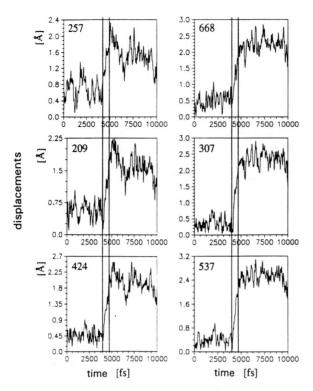

Fig. 4.16. Successive displacements of six Fe atoms within a displacement chain of $Fe_{90}Zr_{10}$ at $T = 1655\,K$ [4.45]

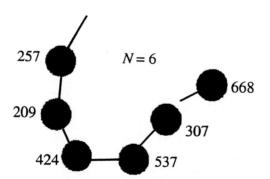

Fig. 4.17. Projection of the displacement chain of Fig. 4.16. Average displacement $\langle R \rangle = 6.4\,\text{Å}$

roomed by small displacements of a cluster of atoms so that the displacement chain stops. Whereas in $Fe_{91}Zr_9$ only Fe atoms are involved in displacement chains and the large Zr atoms remain rather immobile, in $Fe_{24}Zr_{76}$, both, Fe as well as Zr atoms, are involved in the displacement chains which are started and finished by Fe atoms. These results agree with theoretical investigations of Schober et al. [4.49, 50] on vibrations in glassy structures. There are then found collective approximately one-dimensional motions of clusters of atoms.

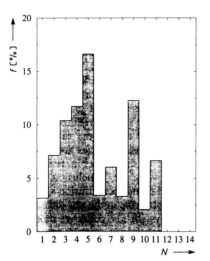

Fig. 4.18. Distribution function $f(N)$ of displacement chains of $N = 1, 2, 3 \ldots$ atoms in relaxed $Fe_{90}Zr_{10}$ at 1655 K taking place within a time interval of 100 ps [4.13, 45]

It is of interest to note that the displacement chains are not related to the formation of quasi-vacancies or quasi-interstitials. If these are created artificially, the MDS shows that they vanish by being smeared out over a cluster of atoms. The relaxation time for this process is of the order of 10^{-13}–10^{-12} s. The displacement chains are not isolated units in the random walk of atoms but take place rather oftenly. Figure 4.18 shows the fraction $f(N)$ contributing to displacement chains of N atoms within a time interval of 100 ps. Summing up the squares of the displacements of all atoms within the chains gives a value of 1067 Å. If the same is done for all atoms a value of 982 Å is obtained. These similar values support the conclusion that the displacement chains govern the diffusion process.

4.4 Diffusion of Hydrogen Isotopes and Light Particles in Amorphous Alloys

In amorphous alloys hydrogen isotopes occupy "interstitial sites" as in crystalline materials. In thermodynamic equilibrium, the hydrogen isotopes move into the interstitial sites with the lowest site enthalpies, h_i. Since each interstitial site can only pick up one H atom with increasing hydrogen concentration, c_H, higher and higher enthalpy minima of the statistical potential are occupied (Fig. 4.19). This phenomenon can be described by Fermi-Bose statistics where the occupation probability of the interstitial sites with site enthalpy, h_i, is given by

$$f_i = \{1 + \exp[(h_i - \mu_S)/kT]\}^{-1}. \tag{4.48}$$

Here, μ_S denotes the Fermi level up to which the interstitial sites are occupied by H atoms; μ_S is determined by the integral over the occupied sites

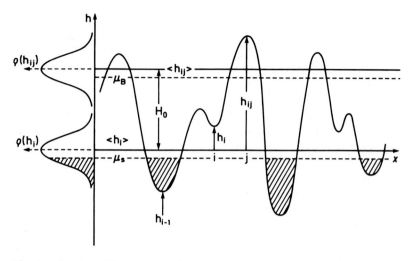

Fig. 4.19. Random diffusion potential for H atoms with occupation of the interstitial sites with small site enthalpies; μ_S and μ_B denote the chemical potentials (Fermi levels) of site and saddle-point enthalpies, respectively

$$c_H = \int_{-\infty}^{\mu_S} \rho(h_i) \, dh_i, \qquad (4.49)$$

where $\rho(h_i)$ corresponds to the distribution function of interstitial site energies.

Assuming for $\rho(h_i)$ a Gaussian distribution with variance h_{RS}^2 and an average site enthalpy $\langle h_i \rangle = h_{0,S}$ the Fermi energy is given by [4.51]

$$\mu_S = h_{0,S} \pm \sqrt{2} h_{RS} \, \mathrm{erf}^{-1} |2c_H - 1|, \qquad \begin{matrix} (+) \text{ for } 2c_H > 1 \\ \nearrow \\ \searrow \\ (-) \text{ for } 2c_H < 1 \end{matrix} \qquad (4.50)$$

where erf^{-1} denotes the inverse error function. Due to the existence of the Fermi level, μ_S (Fig. 4.19), the H diffusion is equivalent to the RBM model. The effective jump frequency v_e for the H atoms may be determined according to the concepts described in Sect. 4.3. Solving the self-consistency equations (4.43) and (4.44), the Fermi level, μ_B, describing the effective jump frequency, $v_e = [v_0/(z/2 - 1)] \exp(-\mu_B/kT)$, is given by

$$\mu_B = h_{0,B} - \sqrt{2} h_{RB} \, \mathrm{erf}^{-1} \left(\frac{4}{z} - \mathrm{erf} \frac{\mu_S}{\sqrt{2} h_{RB}} \right). \qquad (4.51)$$

Rearranging (4.50) by setting $h_{0,S} = 0$ and assuming similar variances $h_{RS} = h_{RB}$ gives

$$\mu_B = h_{0,B} \mp \sqrt{2} h_{RB} \, \mathrm{erf}^{-1} \left| \frac{4}{z} + 2c_H - 1 \right|. \qquad (4.52)$$

In the limit $z \to \infty$, (4.52) gives Kirchheim's result [4.51]:

$$\mu_B = h_{0,B} \mp \sqrt{2} h_{RS} \operatorname{erf}^{-1} |2c_H - 1|. \tag{4.53}$$

Equation (4.53) has been derived previously [4.51], under the assumption of constant barriers, i.e., $h_{RB} \to 0$. The term $h_{0,B}$ in (4.51) corresponds to the average activation enthalpy if the H atoms just occupy half of the interstitial sites ($c_H = 1/2$, $z \to \infty$). The second terms in (4.51) and (4.52) describe the effect of the occupation of interstitial sites with increasing site enthalpies if c_H increases. For $c_H = 0$, the effective activation enthalpy is largest and the minimum value is obtained for $c_H = 1$:

$$\mu_B^{\max}_{\min} = h_{0,B} - h_{RB}\sqrt{2} \times \begin{matrix} \nearrow \operatorname{erf}^{-1} \left|\dfrac{4}{z} - 1\right| \\ \\ \searrow \operatorname{erf}^{-1} \left|\dfrac{4}{z}\right| \end{matrix} \tag{4.54}$$

The negative term in the parentheses of (4.51) expresses the fact that in a three-dimensional random walk the diffusing atoms may avoid the high-energy barriers and since μ_S increases with increasing H content, the effective activation enthalpy decreases.'

The predicted decrease of the effective activation enthalpy with increasing hydrogen content has been investigated in ferromagnetic amorphous alloys by means of magnetic relaxation measurements. This method is based on measurements of isothermal relaxation curves of the initial susceptibility, $\chi(t, T)$, as a

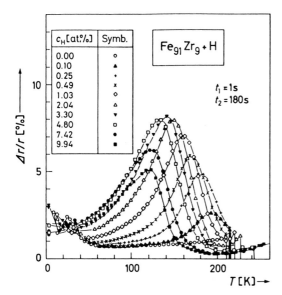

Fig. 4.20. Isochronal relaxation spectra of hydrogen in amorphous $Fe_{91}Zr_9$ for different amounts of H [4.53]

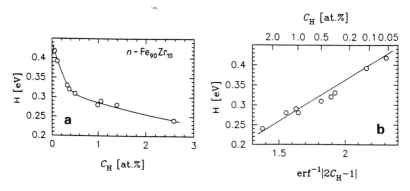

Fig. 4.21. Analysis of activation enthalpy of hydrogen according to (4.50); (a) H vs c_H, (b) H vs $\mathrm{erf}^{-1}|2c_H - 1|$

function of time [4.46]. From these relaxation curves so-called isochronal relaxation spectra of the reluctivity, $r(t, T) = 1/\chi(t, T)$, are constructed according to

$$\Delta r(t_1, t_2) = \frac{1}{\chi(t_2, T)} - \frac{1}{\chi(t_1, T)}. \tag{4.55}$$

Figure 4.20 shows a series of relaxation spectra of a FeZr amorphous alloy for different amounts of hydrogen content [4.53, 54]. The shift of the relaxation maximum to lower temperatures indicates the decrease of the effective activation enthalpy with increasing hydrogen content. Figure 4.21 gives a comparison of the measured effective activation enthalpies with (4.50). The parameters found are $h_{0,B} - h_{0,S} = 0.46\,\mathrm{eV}$ and $h_{RB} = 0.13\,\mathrm{eV}$. These values are reasonable in such a sense that they are similar to the activation parameters found in crystalline disordered alloys. It is of interest to note that μ_S varies between 0.46 and 0.24 eV for a variation of c_H from 0.1 to 2.6 at.%.

4.5 Magnetic After-Effects and Induced Anisotropies Due to Double-Well Systems in Amorphous Alloys

Whereas in diffusion experiments the tracer atoms perform long-range migrations, MAE measurements detect local rearrangements of mobile atomic units only. Therefore, MAEs are usually observed at much lower temperatures than the long-range diffusion. MAEs are most directly observed by time-dependent changes of the characteristic properties of the hysteresis loop, e.g., by a decrease of the initial susceptibility $\chi(t, T)$ or by an increase of the reluctivity $r(t, T) = 1/\chi(t, T)$ and of the coercivity, H_C, as shown in Fig. 4.22 for the magnetically soft amorphous alloy $Co_{58}Ni_{10}Fe_5Si_{11}B_{16}$. Figure 4.23 shows isothermal relaxation curves of $r(t, T)$ as measured for an amorphous $Fe_{80}B_{20}$ alloy [4.55]. The isochronal relaxation curves then may be constructed from the measurements of

4.5 Magnetic After-Effects and Induced Anisotropies Due to Double-Well Systems   119

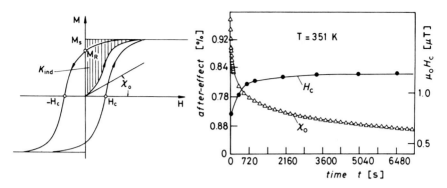

Fig. 4.22. Hysteresis loop (*left*) and isothermal relaxation curves of the initial susceptibility, χ_0, and of the coercivity, H_c, of amorphous $Co_{58}Ni_{10}Fe_5Si_{11}B_{16}$ at 315 K after pre-annealing at 480 K for 220 min

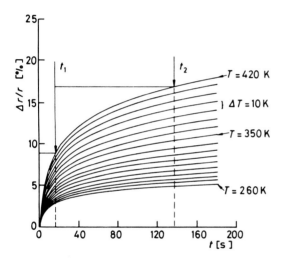

Fig. 4.23. Isothermal relaxation curves of the reluctivity $r(t) = 1/\chi(t)$ of $Fe_{80}B_{20}$ after pre-annealing at 480 K for 11 min

Fig. 4.23 according to (4.55). The quantitative analysis of the relaxation maxima is based on superposed exponential relaxation processes

$$r(t, T) = r_0 + r_\infty \int_0^\infty \{[1 - \exp(-t/\tau)]p(\tau)\}\,d\tau, \tag{4.56}$$

where $r_0 - r_\infty$ denotes the total relaxation amplitude and $p(\tau)$ the probability distribution function of relaxation times $\tau = \tau_0 \exp(H/kT)$; r_∞ decreases according to a $1/T$ law with increasing temperature. From a fit of (4.56) to the isothermes of Fig. 4.23, the spectrum of activation enthalpies is obtained as shown in Fig. 4.24. The steep decrease of the distribution function $p(H)$ on the high-enthalpy side indicates that due to the finite measuring time of $t_2 = 180$ s the large relaxation times cannot be detected. Therefore, increasing the measuring time t_2 gives more information on the larger activation enthalpies as

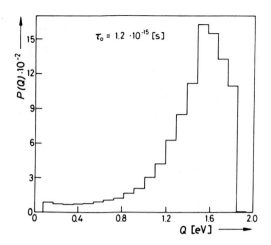

Fig. 4.24. Spectrum of activation enthalpies as obtained from Fig. 4.23 for $Fe_{80}B_{20}$ using (4.55)

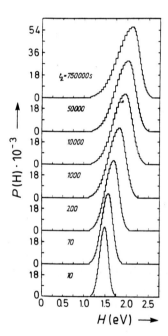

Fig. 4.25. Spectra of activation enthalpies obtained for $t_1 = 1\,s$ and different measuring times t_2 from isothermal relaxation curves of $Co_{58}Ni_{10}Fe_5Si_{11}B_{16}$. The increase of the effective activation enthalpy with increasing t_2 is demonstrated [4.56]

shown in Fig. 4.25, where we present the distribution functions $p(H)$ obtained for increasing measuring times t_2 for an amorphous $Co_{58}Ni_{10}Fe_5Si_{11}B_{16}$ alloy [4.56]. Comparing the results for $t_2 = 10\,s$ and $t_2 = 750,000\,s$ the average activation enthalpy increases from 1.47 to 2.03 eV. Figure 4.26 reviews the activation enthalpies as obtained for an amorphous $Co_{58}Ni_{10}Fe_5Si_{11}B_{16}$ alloy by different regimes of measuring times and measuring temperatures these different

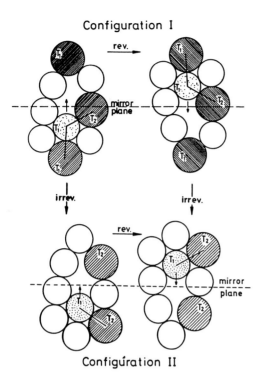

Fig. 4.26. Planar hard sphere model of a double-well system showing an isomorphic reorientation of a pair axis (reversible process) and a polymorphic transition between configurations I and II (irreversible process)

techniques monitor different parts of the activation enthalpy spectra of diffusion and structural relaxation.

The abrupt cut-off observed in the case of the MAE and the induced anisotropy on the high-enthalpy side again is due to the finiteness of the measuring times. It is obvious that the long-range and long-time self-diffusion enthalpies correspond to the upper limit of the spectrum of activation enthalpies.

The magnetic relaxation processes, MAE and induced anisotropy, are adequately described by so-called Double-Well Systems (DWS) schematically shown in Fig. 4.26 [4.57]. Here, an atom T_1 may occupy two equivalent positions with respect to the surrounding nearest neighbour atoms which is guaranteed by the presence of a local mirror plane. In the upper part of Fig. 4.26 an "isomorphic" reversible transition is shown, and the lower part of Fig. 4.26 shows the change of configuration I into a configuration II by a "polymorphic" irreversible transition. The relaxation processes may be attributed to the reorientation of the atom pair axis of two neighbouring atoms T_1 and T_2. Those orientations of the pair axes are preferred where the magnetic interaction energy of the atom pair is a minimum. According to Fig. 4.27, DWS are characterized

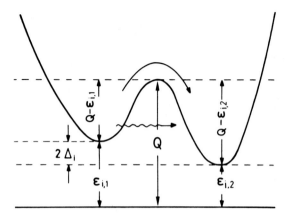

Fig. 4.27. Model of a DWS showing the modification of the potential minima by magnetic interaction energies [4.57]

by their activation enthalpies, H, for the pair-axis reorientation and the splitting energy

$$2\Delta = \varepsilon_1 - \varepsilon_2 \qquad (4.57)$$

between the two neighbouring energy wells. The energies $\varepsilon_{i,j}$ contain the contributions of the magnetic interactions of the atom pair, but also a structural term corresponding to atomic rearrangements as, e.g., the transition from configuration I into configuration II.

The MAE of the reversible susceptibility, $\chi(t, T)$, is due to a decrease of the domain wall mobility due to the orientational rearrangement of atom pairs within the domain wall. Here, the atom pair axes tend to arrange parallel or perpendicular to the spontaneous magnetization depending on the angular dependence of ε_i. Starting from an isotropic distribution of the orientations of atom-pair axes, these orient into their energetically preferred orientations as schematically shown in Fig. 4.28. This process leads to a decrease of the total domain wall potential with increasing time, thus resulting in a decrease of the susceptibility.

Whereas the MAE is restricted to the rearrangements of DWS within the domain walls, the formation of an Induced Magnetic Anisotropy (IMA) is related to an overall preferred orientation of atom-pair axes in the bulk parallel or perpendicular to the spontaneous magnetization as demonstrated by Fig. 4.29.

The reversibility of the reorientation processes of DWS has been tested by so-called cross-over experiments based on a change of the measuring temperature during two isothermal annealing runs [4.58]. Figure 4.30 shows two isothermal annealing curves of K_{in} for $T = 473$ K and $T = 525$ K. These two isothermes cross each other at $t \approx 2 \times 10^4$ min because, at low temperatures, the increase of K_{in} proceeds with a smaller rate than for higher temperatures, whereas the equilibrium values $K_{in}(\infty)$ for $t \to \infty$ decrease with increasing T

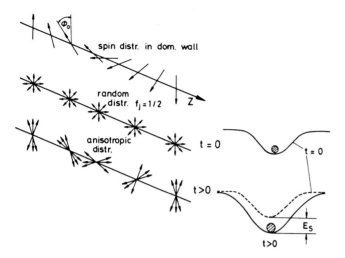

Fig. 4.28. Schematic model of the MAE of the susceptibility. The random distribution of atom pair axes within a domain wall at $t = 0$ changes into a fan-like distribution of pair axes following the rotating orientation of magnetization within the domain wall. The deeper domain wall potential leads to a reduced susceptibility

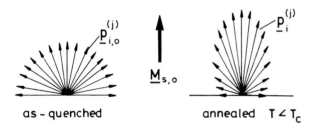

Fig. 4.29. Schematic model of IMA due to a change of the average orientation of atom pair axes of DWSs. Distribution functions, $P_{i,0}$, in the as-quenched state and, P_i, after magnetic annealing

because of the $1/T$ dependence of the saturation amplitude r_∞. After an annealing time of $t = 4.5 \times 10^4$ min at 473 K, K_{in} has approached the equilibrium value K_{in}^∞ for $T = 525$ K. Changing now the annealing temperature to $T = 525$ K leads to a steep decrease of K_{in} within a time interval $< 10^3$ min. At larger annealing times K_{in} again increases now approaching the smaller equilibrium value for $T_A = 525$ K. This inverse annealing behaviour of K_{in} is due to the existence of a spectrum of activation enthalpies. During the annealing time up to 4.5×10^4 min all DWS with small relaxation times τ or low activation enthalpies have approached their thermal equilibrium distribution, whereas those of larger activation enthalpies have not. Increasing now the annealing temperature, those DWS's which are in a thermal equilibrium have to rearrange, i.e., for an increased annealing temperature the orientational order of pair axes

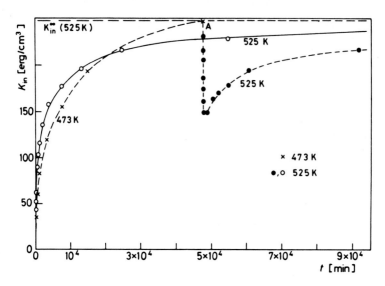

Fig. 4.30. Crossover behaviour of the IMA studied by changing the annealing temperatures from 473 to 525 K at $t = 4.5 \times 10^4$ min. The thermodynamic equilibrium value K_i^∞ for 525 K is given by the *upper broken line* [4.58]

is reduced, leading to a steep decrease of K_{in}. DWS with large activation enthalpies after the temperature change, however, proceed to approach their equilibrium value with a larger rate.

4.6 Viscosity and Internal Friction of Amorphous Alloys

Amorphous solids are subjected to structural changes under the influence of external fields. Due to their metastable structure, amorphous solids relax into more stable configurations if they are exposed to annealing temperatures where thermally activated processes take place within the time window of investigations. Mechanical relaxation phenomena are well known from crystalline solids where these processes are attributed to the displacement of dislocations and the rearrangement of atomic defects. The situation in rapidly quenched alloys is somewhat more complex because here relaxation processes, in general, are due to the superposition of irreversible and reversible microstructural changes due to the presence of free volumes (quasi-vacancies) in supersaturation. Relaxation processes due to DWS, as described in Sect. 4.5 certainly are the source of mechanical relaxation processes far below the glass temperature, T_g. These relaxation processes are local, quasi-elastic polarization processes of a small number of atoms, e.g., the reorientation of an atom pair. Near the glass temperature mechanical relaxation processes become more and more like the Newtonian viscous flow in liquids, where the viscosity results from a collective

4.6.1 Viscosity Measurements

The viscosity, η, may be determined by a tensile creep experiment where the tensile stress, σ, and the strain rate, $\dot{\varepsilon}$, are related by the material law

$$\sigma = 3\eta(t, T)\dot{\varepsilon}. \tag{4.58}$$

Equation (4.58) can be used either to determine η for constant σ or constant $\dot{\varepsilon}$ at a certain temperature, T. Extensive experimental results are due to *Spaepen* et al. [4.10, 59] for amorphous $Pd_{82}Si_{18}$, $Pd_{77.5}Cu_6Si_{16.5}$, $Pd_{40}Ni_{40}P_{19}Si_1$ and to *Chen* for $Pd_{40}Ni_{40}P_{20}$ and $Pd_{48}Ni_{32}P_{20}$ [4.60].

At temperatures far below the glass temperature the structural relaxation leads to a linear increase of η with time. As an example, Fig. 4.31 shows isothermal relaxation curves for $Pd_{82}Si_{18}$ [4.10]. After the measurement of each isothermal relaxation curve the viscosity was measured at several, somewhat lower temperatures, i.e., in a so-called isoconfigurational state. As in the case of self-diffusion, it could be shown that η after a pretreatment at higher temperatures behaves as a reversible material property. This also holds for temperatures around the glass temperature, where it could be shown for $Pd_{40}Ni_{40}P_{19}Si_1$ that after large annealing times the viscosity approaches its isoconfigurational equi-

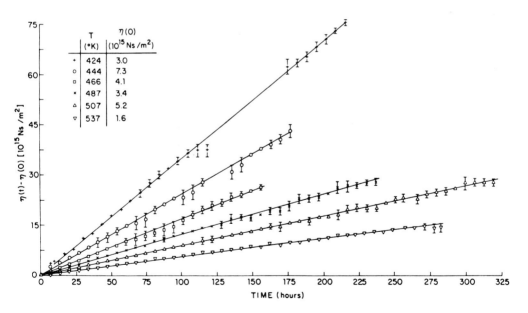

Fig. 4.31. Viscosity $\eta(t)$ of amorphous $Pd_{82}Si_{18}$ as a function of time for different measuring temperatures [4.10]. The table lists the initial viscosity $\eta(0)$

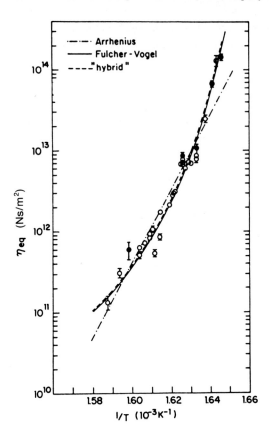

Fig. 4.32. Analysis of the temperature dependence of $\eta(T)$ in the relaxed, isoconfigurational state. The best fit is given by $\eta(T) = 1.2 \times 10^8 \exp[338/(T-583)]$ J Pa S [4.10]

librium value independent on the type of pretreatment [4.59]. The temperature dependence of η in the isoconfigurational state far below the glass temperature obeys an Arrhenius law, $\eta(T) = \eta_0 \exp(H/kT)$, with an activation enthalpy of $H = 1.9$ eV for $Pd_{82}Si_{18}$ [4.10].

At temperatures near the glass temperature ($T_g \approx 640$ K for $Pd_{82}Si_{18}$) η obeys the so-called Vogel-Fulcher equation (Fig. 4.32)

$$\eta(T) = \eta_0 e^{A/(T-T_0)}. \tag{4.59}$$

The test with an Arrhenius equation is less satisfactory and a combination of Vogel-Fulcher and Arrhenius equations does not improve the fit [4.59]. For amorphous $Pd_{77.5}Cu_6Si_{16.5}$ *Spaepen* et al. [4.59] found $\eta_0 = 1.2 \times 10^8$ Ns/m², $A = 338$ K and $T_0 = 583$ K.

The non-Arrhenian behaviour of $\eta(T)$ near T_g has been related to the temperature dependence of the free volume, v_f, which according to *Cohen* and *Turnbull* [4.62] is given by

$$v_f(T) = \alpha V_M(T - T_0), \tag{4.60}$$

where α is the coefficient of thermal expansion, V_M the molecular volume. Inserting (4.60) into (4.59) gives the result derived by *Spaepen* et al. [4.10, 59]

$$\eta(T) = \eta_0(T) e^{\gamma v^*/v_f}, \qquad (4.61)$$

where γ is the geometrical factor between 1/2 and 1, v^* the minimum local free volume required for an atomic jump. Equation (4.61) also should hold for the isoconfigurational viscosity far below T_g where no irreversible processes take place, and v_f is temperature independent.

4.6.2 Internal Friction Measurements

Internal friction measurements, in general, are performed in an inverted torsional pendulum on cylindrical or narrow ribbons of amorphous alloys. In the case of a freely oscillating system, due to the internal friction, the amplitude of the oscillation decreases with time. The internal friction, Q^{-1}, then is defined as the relative energy loss $\Delta W/W$ per period multiplied by $1/2\pi$. Another definition is related to the ratio A_i/A_{i+1} of two successive maximum amplitudes. Thus, Q^{-1} is given by [4.63]

$$Q^{-1} = \frac{1}{2\pi}\frac{\Delta W}{W} = \frac{1}{\pi}\ln(A_i/A_{i+1}) = \frac{1}{\pi}\Lambda, \qquad (4.62)$$

where Λ corresponds to the logarithmic decrement. If we deal with a unique Debye process, characterized by a relaxation time, τ, the internal friction of a periodic oscillation with frequency ω is given by

$$Q^{-1}(w,\tau) = \Delta \frac{\omega\tau}{1+\omega^2\tau^2}, \qquad (4.63)$$

where Δ corresponds to the relaxation strength defined as

$$\Delta = \delta M/M_R, \qquad (4.64)$$

where $\delta M = M_U - M_R$ denotes the difference of the unrelaxed and the relaxed elastic modules, which may be defined by the relaxation of the elastic stress $\sigma(t)$ if the specimen is deformed spontaneously at $t = 0$ by a constant strain ε_0. In this case, $\sigma(t)$ relaxes according to

$$\sigma(t) = \varepsilon_0[M_R + \delta M \exp(-t/\tau)]. \qquad (4.65)$$

Equation (4.65) can be interpreted by Hooke's law with a time-dependent elastic modul $M(t) = M_R + \delta M \exp(-t/\tau)$. We now obtain from (4.65) $M_U = M(0) = M_R + \delta M$, and $M_R = M(\infty)$. In the case of a periodically oscillating system, we can define a so-called modulus defect

$$\frac{\delta M}{M_R} = \Delta \frac{1}{1+\omega^2\tau^2}. \qquad (4.66)$$

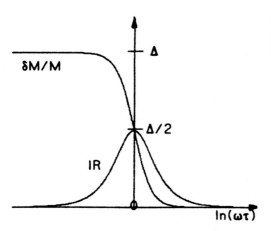

Fig. 4.33. Internal friction, Q^{-1}, and modulus defect, $\delta M/M_R = \Delta$, for a Debye process as a function of $\ln(\omega\tau)$

Comparing (4.63) and (4.66) in Fig. 4.33 it turns out that at $\omega\tau = 1$, the relations

$$Q^{-1}_{max} = \frac{\Delta}{2}, \quad (\delta M/M_R)_{max} = \Delta. \tag{4.67}$$

hold. Accordingly, the maximum internal friction corresponds just to half of the maximum modulus effect (Fig. 4.33).

Internal friction measurements have been performed on PdNiP alloys. As an example, Fig. 4.34 shows Q^{-1} and $\delta M/M_R$ as a function of temperature for a frequency of $\omega = 2.45\,\text{s}^{-1}$. The specimen has been produced by the melt-spin

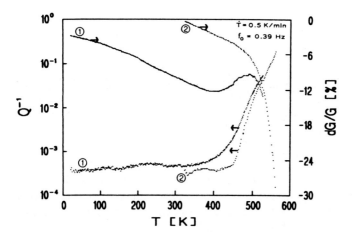

Fig. 4.34. Internal friction, Q^{-1}, and modulus defect, $\delta G/G$ of the shear modulus, for an as-quenched $Pd_{40}Ni_{35}P_{25}$ specimen. ①: first annealing run up to 528 K; ②: second annealing run up to 573 K ($\dot{T} = 0.5\,\text{K/min}, f_0 = 0.39\,\text{Hz}$)

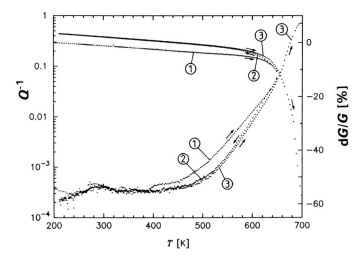

Fig. 4.35. Internal friction, Q^{-1}, and modulus defect, $\delta G/G$ of the shear modulus, for an $Zr_{65}Cu_{17.5}Ni_{10}Al_{7.5}$ alloy [4.65]. ①: first annealing run up to 550 K; ②: decreasing annealing temperatures; ③: annealing run up to 700 K ($\dot{T} = 2$ K/min, $f_0 = 1.33$ Hz)

technique and run ① up to 528 K corresponds to the as-quenched sample. In a second run ②, the Q^{-1} spectrum was measured up to 573 K. In both runs above 400 K, the internal friction increases significantly, whereas the modulus defect decreases monotonically in run ① up to 400 K, where $\delta M/M_R$ again increases indicating an elastic stiffening of the sample. Below 400 K, the internal friction may be attributed to the DWS discussed in Sect. 4.5. In the case of run ②, the modulus decreases continuously with a drastic decrease above 550 K which announces the approach to the melting point. The steep increase of Q^{-1} above 400 K indicates the transition from a predominantly inelastic to a viscoelastic behaviour. The glass temperature may be defined as the crosspoint of the linearly extrapolated internal friction spectrum below and above 400 K. The stiffening of the modulus above 400 K may be attributed to the annealing of excess-free volume quenched-in during the rapid quenching process. Above 500 K, the modulus again decreases due to the thermally activated formation of free volumes and the approach to the melting point.

Internal friction measurements also have been performed on $Zr_{65}Cu_{17.5} \cdot Ni_{10}Al_{7.5}$ [4.65] alloys which are characterized by a large temperature range of $\Delta T = 120$ K between the glass temperature $T_g = 630$ K and the crystallization temperature $T_K = 750$ K [4.60, 64]. Accordingly, the viscoelastic region can be investigated over a broad temperature range. As in the case of the PdNiP alloys, the internal friction increased and the modulus defect shows a drastic decrease above the glass temperature (Fig. 4.35). The onset of crystallization leads to a stiffening of the material giving rise to a reduction of the modulus defect.

Appendix*: Microsectioning by Ion-Beam Sputtering – A Powerful Method to Determine Diffusion Profiles

Investigations of self-diffusion have led to a deeper understanding of diffusion mechanisms not only within crystalline materials, but also within amorphous metallic alloys. For a direct measurement of self-diffusion, the so-called radiotracer technique is a well-suited method. With this method, a thin layer of radio-active atoms is brought on the highly smoothed surface of a specimen, either electrolytically or by evaporation. After diffusion annealing, layer by layer will be leveled down from the specimen. Then, the concentration of the radioactive atoms in each layer is measured. Knowing the thickness of each layer, a depth profile of the tracer concentration can be obtained, which allows the determination of the diffusion coefficient.

The application of the radiotracer technique to amorphous metallic alloys presents the following problems. The upper limit of the temperature interval within which the investigations can be carried out is given by the beginning of crystallization. In order to obtain diffusion data within a sufficiently large temperature interval, as a consequence, very small diffusion coefficients ($< 10^{-20}$ m^2/s) have to be measured. This difficulty can only partly be avoided by long diffusion times, because those are limited by the sometimes fast radioactive decay of the tracer atoms. It follows that, to measure diffusion coefficients within a large temperature interval, a specific sectioning method is required which allows a microsectioning in the nanometer regime. Therefore, conventional mechanical sectioning methods have to be ruled out. Also chemical and electro-chemical methods can only be used with pure metals, because the etching rates of the components of an alloy usually are different.

A well-proved method for the determination of very small diffusion coefficients is microsectioning by ion-beam sputtering [4.66]. If a beam of energetic particles is projected at a target, the particles transfer their energy to the target atoms within a few atomic layers. Within the damage sequence the transferred energy gives rise to many complex processes. One of the basic mechanisms involves a reversal of the incident momentum carried by the ions so that target atoms are ejected from the surface. The process of atomic ejection during ion bombardment is called "sputtering".

A schematic representation of an ion-beam sputtering apparatus is shown in Fig. 4.36. In this apparatus, argon ions are accelerated with an energy of 0.5 to 1 keV onto the specimen. As an ion source, a commercial sputter gun, as used for cleaning surfaces, can be applied. For the determination of diffusion profiles, a highly homogeneous sputter rate is required. This can be achieved either by the application of a strongly defocused ion beam in connection with a turning of the sample, or by scanning the ion beam. In the latter case the elimination of

---

* Prepared by P. Scharwaechter

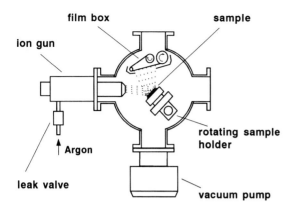

Fig. 4.36. Schematic of an ion-beam sputtering apparatus

neutral atoms is required. The sputtered-off material is collected on a plastic film which is winded on step by step in a very similar way as a film is transported in a camera. The shape of the diffusion profile is determined by subsequently measuring the radio activity of the individual film sections. Knowing the face and the density of the specimen and under the prerequisite of a constant sputter rate the depth of the profile and therefore the diffusion coefficient can be determined from the loss of weight of the specimen. By using this method, diffusion over a length of only a few 10 nm can be measured, and diffusion coefficients down to 10^{-24} m^2/s were determined [4.67].

As a very recent result of self-diffusion measurements Fig. 4.37 shows the Arrhenius plots of self-diffusion for Pd and Au in the quasicrystalline material $Al_{70}Pd_{21}Mn_9$. In contrast to amorphous alloys the Arrhenius plots show a

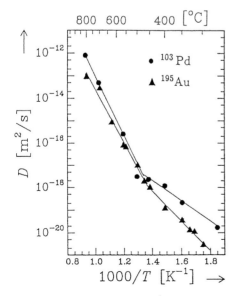

Fig. 4.37. Arrhenius plot of the radio tracer diffusion coefficients of Pd and Au in quasicrystalline AlPdMn

change of slopes at 450 °C, thus indicating a change of the diffusion mechanism [4.68]. It is proposed to interpret the self-diffusion for $T > 450$ °C and activation enthalpies > 2 eV by a vacancy mechanism, whereas the diffusion below 450 °C with activation enthalpies around 1 eV is attributed to a phason-assisted direct diffusion mechanism [4.68]. It should be noted here that the observed change of slopes requires a measuring technique of high precision so far available only by the high-resolution microsectioning technique.

References

4.1 K. Maier, M. Peo, B. Salle, H.-E. Schaefer, A. Seeger: Philos. Mag. A **40**, 701 (1979)
4.2 A. Seeger: In DIMAT-92, ed. by M. Koiwa, K. Hirano, H. Nakajima, T. Okada, Defect and Diffusion Forum **95–98**, 147 (1993)
4.3 A. Seeger, K.P. Chik: Phys. Status Solidi **29**, 455 (1968)
4.4 U. Breier, W. Frank, C. Elsässer, M. Fähnle, A. Seeger: Phys. Rev. B **50**, 5928 (1994)
4.5 R.W. Cahn: 2nd Int'l Workshop on Noncrystalline Solids (San Sebastian 1989)
4.6 T. Egami: In *Amorphous Metals and Semiconductors*, ed. by Haasen, R.I. Jaffee (Pergamon, Oxford 1986) p. 222
4.7 T. Egami, Y. Waseda: J. Non-Cryst. Solids **64**, 113 (1984)
4.8 W.L. Johnson: Mater. Sci. Eng. **97**, 1 (1988)
   W.L. Johnson, A. Peker: In *Science and Technology of Rapid Solidification and Processing*, ed. by M.A. Otooni (Kluwer, Dordrecht 1995) p. 25
4.9 B. Cantor: In [Ref. 4.6, p. 108]
4.10 A.I. Taub, F. Spaepen: Acta Met. **28**, 1781 (1980)
4.11 H. Kronmüller, W. Frank: Rad Eff. Def. Solids **108**, 81 (1991)
4.12 H. Kronmüller, W. Frank, A. Hörner: Mater. Sci. Eng. A **133**, 410 (1991)
4.13 W. Frank, A. Hörner, P. Scharwaechter, H. Kronmüller: Mater. Sci. Eng. A **179/180**, 36 (1994)
4.14 F. Faupel: Phys. Status Solidi (a) **134**, 9 (1992)
4.15 J. Crank: *The Mathematics of Diffusion*, 2nd edn. (Clarendon, Oxford 1975)
4.16 Th. Heumann: *Diffusion in Metallen* (Springer, Berlin, Heidelberg 1992)
4.17 H. Horvath, J. Ott, K. Pfahler, W. Ulfert: J. Mater. Sci. Eng. **97**, 409 (1988)
4.18 J. Horvath, H. Mehrer: Cryst. Latt. Def. Amorph. Mater. **13**, 1 (1986)
4.19 K. Pfahler, J. Horvath, W. Frank, H. Mehrer: In *Rapidly Quenched Metals*, ed. by S. Steeb, H. Warlimont (Elsevier, Amsterdam 1985) p. 755
4.20 K. Pfahler, J. Horvath, W. Frank: Cryst. Latt. def. Amorph. Mater. **17**, 249 (1987)
4.21 J. Horvath, K. Pfahler, W. Ulfert: In *Rapidly Quenched Metals*, ed. by R.W. Cochrane, J.O. Ström-Olsen (Elsevier, New York 1988) p. 409
4.22 P. Scharwaechter, W. Frank, H. Kronmüller: Z. Metallkunde **87**, 885 (1996)
4.23 R. Sizmann: J. Nucl. Mater. **69–70**, 386 (1968)
4.24 R.S. Averback, H. Hahn: Mater. Sci. Eng. Forum **37**, 245 (1989)
4.25 A.K. Tyagi, M.-P. Macht, V. Naundorf: J. Nucl. Mater. **179–181**, 1026 (1991)
4.26 P. Scharwaechter, W. Frank, H. Kronmüller: Z. Metallkunde **87**, 892 (1996)
4.27 C. Rank: Untersuchungen von Diffusions- und Relaxationsvorgängen in metallischen Gläsern durch Kurzzeit-Tempern in einem Spiegelofen. Dissertation, University of Stuttgart (1992)
4.28 W. Frank, J. Horvath, H. Kronmüller: Mater. Sci. Eng. **97**, 415 (1988)
4.29 H. Mehrer, A. Seeger; Crys. Lattice Def. **3**, 1 (1972)
4.30 R.M. Emrick, P.B. McArdle: Phys. Rev. **168**, 1156 (1969)
4.31 F. Faupel, P.W. Hüppe, U. Rätzke: Phys. Rev. Lett. **65**, 1219 (1990)

4.32 U. Rätzke, F. Faupel: Phys. Rev. B **45**, 7459 (1992)
4.33 K. Rätzke, F. Faupel: J. Non-Cryst. Sol. **181**, 261 (1995)
4.34 U. Rätzke, F. Faupel: Scripta Metall. **25**, 2233 (1991)
4.35 G. Rein, H. Mehrer: Philos. Mag. A **45**, 467 (1982)
4.36 M. Beyler, Y. Adda: J. Physique **29**, 345 (1968)
4.37 N.H. Nachtrieb, H. Resing, S. Rice: J. Chem. Phys. **31**, 135 (1959)
4.38 K. Rätzke, H. Heesemann, F. Faupel: J. Phys.: Condens. Matter 7, 7663 (1995)
4.39 K. Rätzke, P.W. Hüppe, F. Faupel: Phys. Rev. Lett. **68**, 2347 (1992)
4.40 I. Webmann, J. Klafter: Phys. Rev. B **26**, 5950 (1982)
4.41 J.W. Haus, K.W. Kehr, J.W. Lyklema: Phys. Rev. **25**, 2905 (1982)
4.42 R. Zwanzig: J. Stat. Phys. **28**, 127 (1982)
4.43 J.W. Haus, K.W. Kehr: Phys. Rpts. **150**, 263 (1987)
4.44 Y. Limoge, J.-L. Bocquet: J. Non-Cryst. Solids **117**, 605 (1990)
4.45 A. Hörner: Selbstdiffusion in metallischen Gläsern: Näherung des effektiven Mediums und molekulardynamische Simulation. Dissertation, University of Stuttgart (1993)
4.46 W. Frank, A. Hörner, P. Scharwaechter, H. Kronmüller: Mater. Sci. Eng. A **179/180**, 36 (1994)
4.47 E.H. Brandt: J. Phys., Condensed Matter **1**, 9985 (1989)
4.48 S. Nosé: J. Chem. Phys. **81**, 511 (1984)
4.49 H.R. Schober: Physica A **201**, 14 (1993)
4.50 H.R. Schober, C. Oligschleger: Phys. Rev. B **53**, 11469 (1996)
4.51 R. Kirchheim: Acta metall. **30**, 1069 (1982); Prog. Mater. Sci. **32**, 262 (1988)
4.52 H. Kronmüller: *Nachwirkung in Ferromagnetika* (Springer, Berlin, Heidelberg 1968)
4.53 A. Hofmann, H. Kronmüller: Phys. Status Solidi (a) **104**, 381 and 619 (1987)
4.54 M. Hirscher, S. Zimmer, H. Kronmüller: Z. Phys. Chem. **183**, 51 (1994)
4.55 H. Kronmüller: In [ref. 4.6, p. 259]
4.56 M. Bourrous, H. Kronmüller: Phys. Status Solidi (a) **113**, 169 (1989)
4.57 H. Kronmüller: Phys. Status Solidi (b) **127**, 531 (1985); ibid **118**, 661 (1983); see [Ref, 4.6, p. 259]
4.58 H. Kronmüller, H.-Q. Guo, W. Fernengel, A. Hofmann, N. Moser: Cryst. Latt. Def. Amorph. Mater. **11**, 136 (1985)
4.59 F. Spaepen, S.S. Tsao, T.W. Wu: In [Ref. 4.6, p. 365]
4.60 H.S. Chen: In [Ref. 4.6, p. 126]
4.61 N. Morito: Mater. Sci. Eng. **60**, 261 (1983)
4.62 M.H. Cohen, D. Turnbull: J. Chem. Phys. **31**, 1164 (1959)
4.63 A.S. Nowick: Adv. Phys. **16**, 1 (1967)
4.64 W. Ulfert: Innere Reibung von Wasserstoff in amorphen Metallen. Dissertation, University of Stuttgart (1989)
4.65 W. Ulfert, H. Kronmüller: Proc. Int'l Conf. on Internal Friction and Ultrasonic Attenuation in Solids, Poitiers, France (1996). J. Physique, to be published
4.66 H. Mehrer, K. Maier, G. Hettich, H.J. Mayer, G. Rein: J. Nucl. Mater. **69/70**, 545 (1978)
4.67 J. Horvath, J. Ott, K. Pfahler, W. Ulfert: Mater. Sci. Eng. **97**, 409 (1988)
4.68 R. Blüher, P. Scharwaechter, W. Frank, H. Kronmüller: to be published

5 Mechanical Properties and Behaviour

H. Jones and E.J. Lavernia

The glassy or fine scale crystalline microstructures and extended or novel composition ranges of rapidly solidified materials might be expected to exhibit interesting behaviour under applied mechanical stress. Metallic glasses exhibit flow stress σ_y as high as $E/50$, where E is Young's modulus (Table 5.1), approaching the levels $\approx E/30$ found for perfect dislocation-free single-crystal metallic whiskers. The associated high values of hardness and wear resistance have been put to good use, for example, in recording/replay or read/write heads for audio, video, computer or instrumental recording machinery, where their high electrical resistivity and, for particular compositions, high magnetic permeability and good corrosion resistance are also critical for the application. Plastic flow at stress $\sigma > \sigma_y$ and temperature $T < T_g$, the glass transition temperature, is localized in shear bands giving way to more homogeneous time-dependent flow at high temperatures, allowing hot forming to be carried out with or without inducing crystallization. The corresponding microcrystalline products of rapid solidification can also exhibit ultrahigh strengths and durability because of combinations of very fine matrix grain size, high volume fraction of hard intermetallic precipitates or dispersoid phases, and/or extended concentrations of hardening alloy additions in solid solution in the matrix phase. Such ultrafine dual or multiphase microstructures are also ideal candidates for superplastic forming and/or diffusion bonding at elevated temperature and low stress, while intermetallic dispersoids can be incorporated that exhibit excellent resistance to dissolution or coarsening, so imparting quite exceptional microstructural stability at elevated temperature. Wide ranging reviews of the mechanical performance of rapidly solidified materials have been published by *Gilman* [5.1], by *Davis* [5.2], by *Li* [5.3], by *Rama Rao* and *Radhakrishnan* [5.4], by *Taub* [5.5], by *Das* and *Froes* [5.6] and by *Davis* et al. [5.7]. References [5.1–5.4] are confined to metallic glasses, [5.5] and [5.6] feature microcrystalline alloys while [5.7] embraces both categories.

5.1 Elastic and Anelastic Behaviour

Young's and shear moduli E and G of metallic glasses are typically 20 to 40% lower than in the crystalline state because the atoms in the glass are not con-

Table 5.1. Mechanical properties of some transition metal/metalloid metallic glasses at room temperature. Yield strength values (measured at strain rate $10^{-4} s^{-1}$) are tensile except for $Pd_{64}Ni_{16}P_{20}$ determined in compression [5.7]

Alloy composition [at.%]	H_v [kg/mm²]	σ_y [MPa]	E [GPa]	H_v/σ_y	E/σ_y
$Ni_{36}Fe_{32}Cr_{14}P_{12}B_6$	880	2730	141	3.2	52
$Ni_{49}Fe_{29}P_{14}B_6Si_2$	790	2380	129	3.3	54
$Fe_{80}P_{16}C_3B_1$	835	2440	–	3.4	–
$Fe_{80}P_{12.5}C_{7.5}$	810	2730	140	2.9	51
$Fe_{80}Si_{10}B_{10}$	830	2910	158	2.8	54
$Fe_{80}B_{20}$	1100	3630	166	3.0	46
$Co_{77.5}Si_{12.5}B_{10}$	1140	3580	190	3.1	53
$Pd_{77.5}Cu_6Si_{16.5}$	500	1440	88	3.4	61
$Pd_{64}Ni_{16}P_{20}$	450	1540	92	2.9	60
Mean H_v/σ_y, E/σ_y:				3.1 ± 0.2	54 ± 5

strained by a lattice to make self-similar displacements. Bulk modulus is, however, only a few percent lower in the glassy state reflecting the dense random packing which gives a density only 1 to 2% less than for the crystalline state. Temperature dependences of moduli are similar to those of the crystalline state, except at the absolute zero at which E does not go to zero for the glass. Elastic stiffness of both glassy and crystalline matrices can be increased both by appropriate alloying but more effectively by incorporating a phase or phases of higher stiffness. Moduli then tend to increase with increasing volume fraction of the stiffer phases according to the law of mixtures. Examples include rapidly solidified crystalline aluminium alloys containing as much as 40 vol.% of dispersed transition-metal aluminide or silicide introduced by extended alloying which raises E from the 70 GPa characteristic of aluminium and its conventional wrought alloys to values as high as 100 GPa. Similar increases can also be obtained by blending in ceramic particulate during consolidation of a rapidly solidified conventional alloy particulate to form a metal matrix composite.

Rapidly solidified materials, in common with other materials, exhibit anelastic as well as elastic behaviour. Application of a tensile strain to a metallic glass, for example, can induce an appreciable creep strain ($\approx 1\%$) which is eventually recovered on removal of the stress. Such anelastic effects give a useful insight into atomic movements associated with processes such as structural relaxation and also gives rise to potential applications. An early example was provided by the Pd–Ag–Si glasses which exhibit exceptionally low attenuation of longitudinal sound waves [5.8].

5.2 Plastic Flow and Fracture Behaviour

Ductile crystalline metallic materials at ambient temperatures undergo extensive plastic flow above a well-defined flow stress σ_y, which can be very low or very high depending on the condition of the material. Deformation occurs by crystallographic slip or twinning in deformation bands that spread throughout the material as a result of work hardening, allowing extensive plastic flow in tension as well as in compression or constrained shear. Metallic glasses differ in that the flow stress is never low and plastic flow above it remains concentrated within highly localised shear bands (Fig. 5.1) [5.2a]. While the local displacements in single shear bands can be very large, corresponding to a true shear of 10 or more, work hardening does not take place so the deformation does not spread to neighbouring areas and so macroscopic ductility in tension is low. Operation of multiple shear bands under the constraints of bending, shear or compression, however, results in large macroscopic as well as local ductility. This, for example, allows reductions as large as 40% by rolling or drawing in spite of the very high flow stress and hardness. Li [5.3] has proposed that shear-band propagation involves incomplete transfer or density variation (excess volume) in the metallic glass from one location to another, so that any excess volume left behind can act as a source for further shear-band formation. This is not to say that metallic glasses are invariably ductile. Embrittlement can occur as a result of structural relaxation or contamination and on crystallization.

Ductile crystalline metals fracture in tension at a maximum stress σ_u by a process of progressive nucleation, growth and linkage of voids. Resulting shear

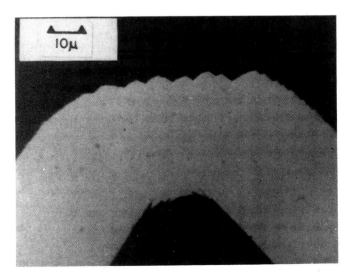

Fig. 5.1. Shear band offsets on the outer surface of a bent metallic glass ribbon [5.2a]

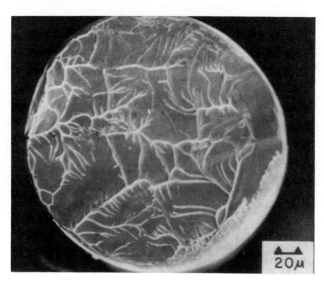

Fig. 5.2. Pattern of local necking protrusions (veins) on oblique fracture surface of $Pd_{77.5}Cu_6Si_{16.5}$ metallic glass wire [5.9]

fracture surfaces are characteristically fibrous derived from the rapid elongation of the remaining unvoided material in the final stages of fracture. The 'vein' pattern (Fig. 5.2) [5.9] characteristic of fracture surfaces of metallic glasses is considered to result from a similar mechanism in which disc-shaped cracks within a shear band expand under stress so concentrating the stress and deformation into decreasing areas between adjacent cracks, these areas finally necking down to failure by shear. The two fracture surfaces thus exhibit a matching array of local necking protrusions (veins) delineating flat areas created by expansion of cracks. This pattern can be readily simulated by tensile separation of two glass microscope slides bonded by a layer of grease. Fracture toughness values for Fe/Ni based metallic glasses range from $\approx 10\,\text{MPa}\sqrt{\text{m}}$ for plane strain failure to $\approx 50\,\text{MPa}\sqrt{\text{m}}$ for tearing failures. Although low, these values are some two orders of magnitude higher than for oxide glasses and are not lower than for high-strength crystalline steels of the same yield strength.

5.3 Strength and Hardness

The highest flow strengths of metallic glasses (up to $3.6\,\text{GPa} \equiv E/45$ for $Fe_{80}B_{20}$) are comparable with the strongest hard-drawn piano wire with corresponding hardness well in excess of $1000\,\text{kg/mm}^2$. Because hardness H is a measure of yield strength under a condition of triaxial constraint, it should be about three times the yield strength in tension, especially for metallic glasses for which no work hardening is involved in the indentation process. Actual results (Table 5.1) [5.7] show $H/\sigma_y \approx 3.2$, in good agreement with expectation. A model which assumes that yield occurs when gliding dislocations cut disclinations spaced about 5 atomic diameters apart predicts $\sigma_y \approx G/22.5$, in reasonable agreement with experimental results [5.7]. Further increments in hardness and strength along with increased stiffness can be obtained by precipitation of nanocrystals within a metallic glass matrix so as to suppress the inhomogeneous shear process that characterizes flow and fracture of the unreinforced glass. For the example in Fig. 5.3 [5.10] tensile fracture strength σ_u of ≈ 900 MPa = $E/55$ is thereby increased to ≈ 1300 MPa with associated increases in H_v, E and elongation to fracture (indicating that partial crystallization of a metallic glass need not be embrittling). Similarly high levels of strength can be retained in such materials by controlled devitrification during consolidation into bulk products. This also has the advantages of increasing E to its crystalline value as well as basing the properties on more stable phases. Results shown in Fig. 5.4 and Table 5.2 [5.11] are for the Al-14wt.%Ni-14.6wt.%Mm (Mm: Mischmetal)

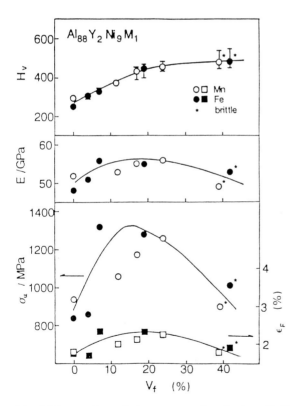

Fig. 5.3. Increase in tensile fracture strength σ_u, hardness H_v, Young's Modulus E and elongation to fracture ε_F in $Al_{88}Y_2Ni_9M_1$ (M = Mn or Fe) metallic glass ribbon, on introduction of a volume fraction V_f of fcc α-Al nanocrystals into the metallic glass matrix [5.10]

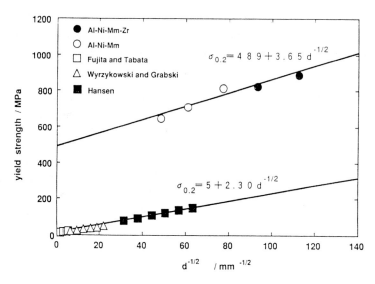

Fig. 5.4. 0.2% proof strength $\sigma_{0.2}$ vs inverse square root of grain size d for rapidly solidified and consolidated Al–Ni–Mm(–Zr) alloys (o, •) compared with results for wrought unalloyed aluminium (□, △, ■) [5.11]

Table 5.2. Tensile fracture strength σ_u, 0.2% proof strength $\sigma_{0.2}$, Young's modulus E and elongation to fracture ε_F as a function of grain size for rapidly solidified Al–Ni–Mm(–Zr) alloys consolidated by extrusion [5.11]

Alloy composition (wt.%)	σ_u [MPa]	$\sigma_{0.2}$ [MPa]	E [GPa]	ε_F [%]	Grain size [nm]
Al-14.9 Ni-3.3 Mm-5.1 Zr	1000	886	97	1.2	80
Al-14.8 Ni-6.6 Mm-2.3 Zr	892	823	96	4.9	120
Al-14.0 Ni-14.6 Mm	909	815	104	0.7	170
Al-14.0 Ni-14.6 Mm	788	711	104	2.4	270
Al-14.0 Ni-14.6 Mm	730	646	103	2.5	420

glass-forming composition and two related Zr-containing compositions, compared with data for unalloyed wrought aluminium. The alloys were rapidly solidified by high-pressure gas atomization to form a mainly amorphous or nanocrystalline powder which was then consolidated by extrusion at relatively high temperature ($\approx 360\,°\text{C} \equiv 0.7\,T_m$, where T_m is the melting temperature). The result was a nanotriplex microstructure of the three or four equilibrium phases, i.e., α-Al solid solution with grain size down to 80 nm stabilized with a high volume fracture of hard intermetallics, Al_3Ni and $Al_{11}Mm_3$ of size ≈ 30 nm and Al_3Zr of size ≈ 15 nm. Figure 5.4 suggests a Hall-Petch (grain size) contribution to σ_y as large as 400 MPa matched by a contribution of some 500 MPa from dispersion and solid solution hardening. A similar proportion of Hall-

5.3 Strength and Hardness   141

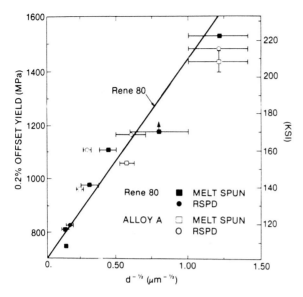

Fig. 5.5. As Fig. 5.4 but for Rene 80 and IN738 Ni-based superalloy rapidly solidified materials. Alloy A = IN738; RSPD = low pressure plasma deposition [5.12]

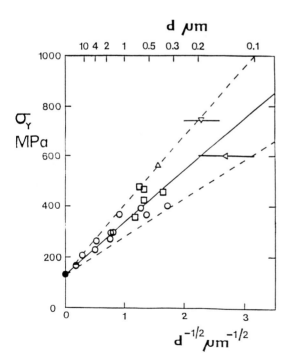

Fig. 5.6. As Fig. 5.4 but for some high-strength magnesium alloy rapidly solidified particulate extrusions (○, □, △, ◁, ▽) compared with Mg-9 wt.% Al-1 wt.% Zn ingot, ● as-cast and heat treated, and ○ extruded [5.13]

Petch contribution is evident for the rapidly solidified nickel-based superalloys in Figure 5.5 [5.12]. An even larger relative Hall-Petch contribution is evident for rapidly solidified Mg–Zn–Al based alloys (Fig. 5.6) [5.13] for which the slope of σ_y vs inverse square root of grain size is $210 \pm 60\,\text{MPa}\,\mu\text{m}^{1/2}$ compared with $115\,\text{MPa}\,\mu\text{m}^{1/2}$ for the aluminium-based alloys in Fig. 5.4 and $770 \pm 150\,\text{MPa}\,\mu\text{m}^{1/2}$ for the nickel-based alloys in Fig. 5.5. The most highly developed of this class of high-strength rapidly solidified light alloys is the 8009 composition (Al-8.5 wt.% Fe-1.3 wt.% V-1.7 wt.% Si) developed by Allied Signal. They used melt-spun ribbon as the starting material which was pulverized, vacuum hot-pressed and then hot-worked by extrusion, rolling or forging. This material contains some 27 vol.% of 50 nm size cubic $Al_{12}(Fe,V)_3Si$ dispersoid in an α-Al solid solution matrix of grain size $\approx 0.1\,\mu\text{m}$. Although strength at normal temperatures does not exceed that of conventional high-strength wrought aluminium alloys, Fig. 5.7 [5.14] shows that the strength at temperatures above $\approx 200\,°\text{C}$ ($\approx 0.5\,T_m$) is much superior, attributable to the unrivalled stability to coarsening of the high volume fraction silicide dispersion. This resistance to coarsening is attributable to the low diffusivity and solubility of iron and vanadium in the α-Al matrix and possibly also to a low interfacial energy of the α-Al/silicide interface. While Fig. 5.7 shows a decrease in strength with increase in test temperature more or less in line with the decrease in modulus, certain ordered intermetallics show an increase in yield strength with increase in temperature. Figure 5.8 [5.15] shows results for Ni_3Al which exhibits this behaviour and it is notable that this effect persists at the higher strength level afforded by rapid solidification combined with 1 at.% added boron, resulting in a four-fold increase in strength compared with the boron-free slowly solidified material. Finally, Fig. 5.9 [5.16] shows that increase in yield strength as a result of processing of silicide strengthened 8009 alloy eventually results in loss of fracture toughness, but that the threshold σ_y for this loss can be substantially increased by ensuring that embrittling aluminide phases do not feature in the final microstructure.

5.4 Fatigue and Wear Behaviour

S–N (stress vs number of cycles to failure) curves for metallic glasses exhibit a well-defined fatigue limit corresponding to $\approx 0.35\,\sigma_y$. Such a limit for crystalline alloys is attributed to strain ageing which is manifest as serrated flow in monotonic deformation, which also occurs for metallic glasses exhibiting a fatigue limit. Fatigue life of metallic glasses above the fatigue limit is, however, normally much shorter than for equivalent crystalline materials of the same strength. This is because work hardening in the crystalline material tends to disperse plastic flow making initiation of fatigue cracks difficult. The absence of work hardening in the glass allows rapid crack initiation to occur at relatively low stresses. Because rates of crack propagation are comparable in the crystal-

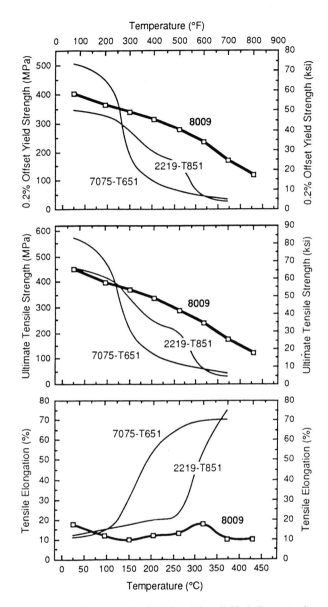

Fig. 5.7. Tensile properties of 8009 rapidly solidified alloy extrusions as a function of test temperature compared with high-strength wrought ingot alloys 2219-T851 and 7075-T651 [5.14]

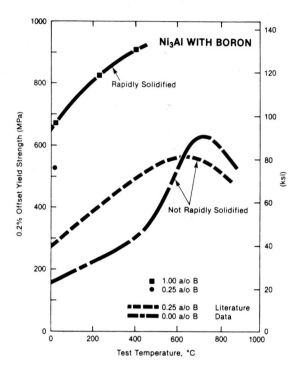

Fig. 5.8. Yield strength vs test temperature for Ni_3Al showing the effects of rapid solidification and of added boron [5.15]

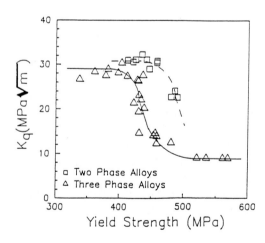

Fig. 5.9. Fracture toughness vs yield strength for rapid-solidification-processed 8009 showing embrittling effect of a third phase (acicular $Al_{13}Fe_4$ or blocky Al_7V) when composition and processing are not sufficiently optimized [5.16]

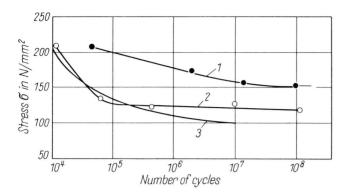

Fig. 5.10. S–N curves for rapidly solidified atomized powder extrusions of Al-8 wt.% Zn-2.5 wt.% Mg-1.0 wt.% Cu-1.5 wt.% Co (*1*) compared with ingot extrusions of 7075-76 (*2*) and commerical wrought 7075-T6510 (*3*) [5.17]

line and glassy materials, the net result is a shorter lifetime and lower fatigue limit for the glassy material. Improved fatigue performance can result from rapid solidification in crystalline materials as illustrated in Fig. 5.10 [5.17] for atomized powder extrusions of Al-8 wt.% Zn-2.5 wt.% Mg-1.0 wt.% Cu-1.5 wt.% Co compared with ingot extrusions of 7075-T6 and commercial wrought 7075-T6510 aluminium alloys.

Although some investigators have found correlations between wear rates of metallic glasses and their hardness, results can be complicated by occurrence of crystallization, e.g., in the wear debris. *Morris* [5.18] found that abrasive wear of two Ni–B–Si metallic glasses was a factor of three or four larger for crystalline steels of the same hardness, whereas the metallic glasses were relatively more wear resistant under sliding conditions. More notable improvements in wear resistance can be achieved by exploiting the capability of rapid solidification to disperse, down to a fine scale, high volume fractions of hard boride, carbide or intermetallic phases in a metallic matrix. Examples include direct injection of ceramic particles into the droplet spray in spray forming, into the melt pool in melt-spinning or into the melt zone in laser surface traverse melting. In the former two cases, the reinforcement is general, whereas in the latter it is confined to the treated surface, enabling wear resistance to be imparted exactly where it is required.

A different approach is to generate a metallic glass precursor by rapidly solidifying a superalloy or tool steel with enhanced boron or carbon contents. A microcrystalline matrix reinforced with a high volume fraction of finely distributed hard boride or carbide then results from hot consolidation above the crystallization temperature, or more directly from the solidification conditions. Such materials have been used as tooling for extrusion and die casting. A Ni–Fe–Mo–B alloy casting die insert made by this route showed no significant erosion or heat checking after 125,000 shots, whereas the typical life of a

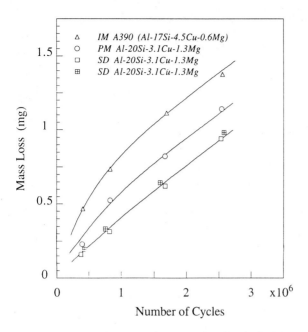

Fig. 5.11. Mass loss in wear testing vs number of cycles for RS-processed Al-20 wt.% Si-3.1 wt.% Cu-1.3 wt.% Mg compared with die-cast A390 alloy (△). *PM* indicates rapidly solidified atomized powder processed, *SD* indicates spray-deposited [5.23]

conventional H13 tool insert under the same conditions was 100,000 shots [5.19]. Similarly, the equilibrium aluminide reinforced Al-14 wt.% Ni-14.6 wt.% Mm alloy described in Sect. 5.3 has been used successfully for higher performance fastener machine parts [5.20] because of its combination of high fatigue strength, low expansion coefficient, good wear resistance and good thermal stability. Finer distributions of primary silicon and aluminide phases in rapidly solidified hypereutectic Al–Si(–Fe)-based compositions result in raised seizure loads when substituted for die cast A390 alloy (Al-17 wt.% Si-4 wt.% Cu-0.6 wt.% Mg) developed for automotive cylinder blocks [5.21]. Development of a near net shape extrusion technology has enabled mass production of rapidly solidified Al-17 Si 5Fe-4.5 Cu-1.5 Mg 0.6 Mn (wt.%) rotors and vanes in an all-aluminium automobile compresser [5.22]. Figure 5.11 [5.23] shows the reduction in metal loss in wear testing of RS-processed Al-20 wt.% Si-3.1 wt.% Cu-1.3 wt.% Mg compared with die cast A390 alloy.

5.5 Creep and Hot Deformation Behaviour

Metallic glasses undergo considerable creep deformation at temperatures close to, but below, T_g, the dependence of minimum creep rate $\dot{\varepsilon}_s$ on stress σ being given by $\dot{\varepsilon}_s = A\sigma^n$ with the n variable between 1, indicating Newtonian viscous flow, and 12, depending on temperature and stress level. For $Pd_{80}Si_{20}$, a transition from Newtonian flow to nonlinear behaviour occurs at a shear stress of about 300 MPa at 500 K. Compared with crystalline metals and ceramics, Newtonian viscous flow is observed in metallic glasses at much lower temperatures and up to much higher stresses. Creep rupture data for $Fe_{40}Ni_{40}P_{14}B_6$ glass in Fig. 5.12 [5.4] conforms to the relationship $\dot{\varepsilon}_s \propto t_r^{-0.8}$, where t_r is rupture life, similar to the Monkman-Grant relationship $\dot{\varepsilon}_s \propto t_r^{-1}$ found for crystalline materials. Minimum creep rate $\dot{\varepsilon}_s$ can be correlated with temperature T and applied stress σ in this instance by means of the empirical relationship $\log \sigma = 5.36 + 0.137(\log \dot{\varepsilon}_s - 0.034 T)$, as shown in Fig. 5.13 [5.4]. *Davis* et al. [5.7] pointed out that the homogeneous flow behaviour of metallic glasses at temperatures approaching T_g could be applied with advantage to hot-work them into useful shapes. Thus, a small gear has reportedly been microforged from 3 mm diameter $La_{55}Al_{25}Ni_{20}$ glass at 30 MPa and 500 K [5.24] following demonstration that such material can be drawn from 2.5 mm to 25 μm diameter at temperature 485 K, strain rate 0.05 s^{-1} and a drawing stress of 15 MPa [5.25] and blow

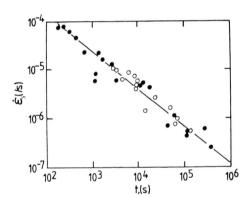

Fig. 5.12. Correlation between minimum creep rate $\dot{\varepsilon}_s$ and rupture time t for $Fe_{40}Ni_{40}P_{14}B_6$ glass (stress range 600 to 1600 MPa, temperatures o 523 K ● 573 K). From [5.4], with data from *Gibeling* and *Nix*: Scr. Metal. **12**, 919 (1978)

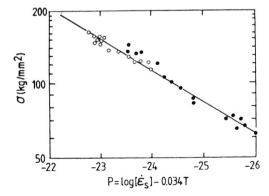

Fig. 5.13. Correlation between minimum creep rate ε_s in s^{-1}, applied stress σ and test temperature T (in K) for $Fe_{40}Ni_{40}P_{14}B_6$ glass. From [5.4], with data from *Gibeling* and *Nix* as for Fig. 5.10

formed at 473 K at differential pressure 0.1 MPa in less than 2 s [5.26] while retaining the glassy structure.

Creep behaviour of crystalline rapidly solidified materials is determined by the combination of ultrafine matrix grain size and the presence of volume fractions of hard dispersed phases. The ultrafine grain size on its own would be expected to enhance the rate of diffusional creep processes at low stresses but the presence of hard dispersoids tends to counter this enhancement and, indeed,

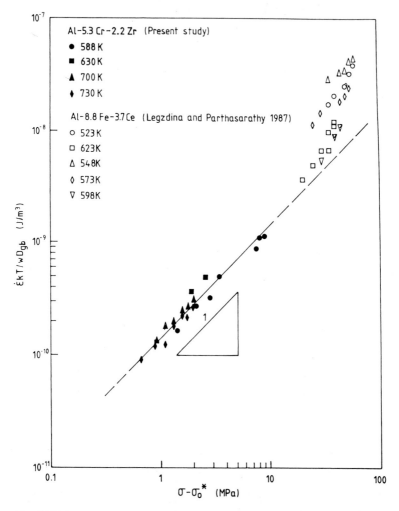

Fig. 5.14. Temperature-compensated creep rate vs effective stress for rapid solidification processed Al-5.3 wt.% Cr-2.2 wt.% Zr and Al-8.8 wt.% Fe-3.7 wt.% Ce. Here, $\dot{\varepsilon}$ is the minimum creep rate, k is Boltzmann's constant, T is the test temperature, w the grain boundary width, D_{gb} is grain boundary self-diffusivity, σ is applied stress and σ_0^* is operative threshold stress. In the absence of a measurement σ_0^* has been taken as zero for the Al–Fe–Ce data. For Al–Cr–Zr, σ_0^* was determined to be in the range 0.25 to 1.3 MPa for the temperature range investigated [5.27]

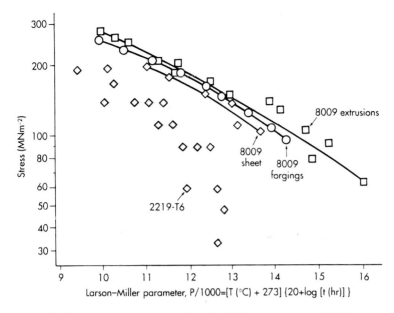

Fig. 5.15. Creep rupture behaviour of rapid solidification processed 8009 aluminium alloy sheet extrusion and forgings in the temperature range 200 to 480 °C (0.5 to 0.8 T_m) compared with wrought ingot high temperature 2219-T851 aluminum alloy [5.14]

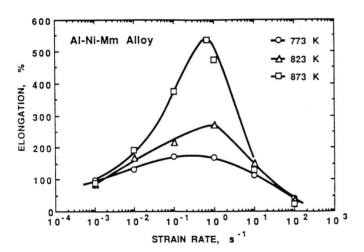

Fig. 5.16. Elongation to fracture vs strain rate for RS-processed Al-14 wt.% Ni-14 wt.% Mm nanocrystalline alloy tested in tension at 500, 550 and 600 °C [5.30]

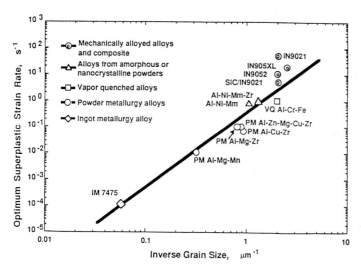

Fig. 5.17. Optimum strain rate for superplasticity vs grain size for high-strength aluminium alloy materials produced by different routes: ◇ wrought ingot processed; ○, △ rapidly solidified atomized powder processed, □ vapour quenched, ⊙ mechanically alloyed [5.31]

can raise the stress exponent n in $\dot{\varepsilon} = A\sigma^n$ to very high values. Results for rapid solidification-processed Al-5.3 wt.% Cr-2.2 wt.% Zr and Al-8.8 wt.% Fe-3.7 wt.% Ce in Fig. 5.14 [5.27] indicate Newtonian behaviour ($n = 1$) with a small threshold stress σ_0^* of less than 1 MPa with indications of a transition to power-law creep at $\sigma - \sigma_0^* \geq 10$ MPa. The measured activation energy for creep was consistent with grain-boundary diffusion being the controlling mechanism although the diffusivities involved are six orders of magnitude less than expected for pure aluminium, doubtless a consequence of the high volume fraction of dispersed aluminides present. Creep rupture performance of rapid solidification processed 8009 sheet, extrusions and forgings, in which a high volume fraction of silicide dispersoid is present, is markedly superior to that of the best available high-temperature wrought aluminium alloy 2219-T6 (Fig. 5.15) [5.14].

The ultrafine grain structure so typical of rapidly solidified crystalline materials could be expected to give rise to superplastic behaviour under certain conditions. Examples include a rapid-solidification-processed white cast iron (Fe-3.0 wt.% C-1.5 wt.% Cr) giving 940% elongation to fracture tested in tension at strain rate 1.7×10^{-4} s^{-1} and temperature 700 °C (0.67 T_m) [5.28] and rapid solidification processed Mg-9 wt.% Al-1 wt.% Zn reportedly giving elongations to fracture exceeding 1000% when tested at strain rate 1.7×10^{-3} s^{-1} and temperature 275 or 300 °C ($\approx 0.6\, T_m$) [5.29]. Both materials had matrix grain sizes ≈ 1 μm stabilized by a distribution of carbide and aluminide respectively. Figure 5.16 [5.30] shows results for the rapid-solidification-processed Al-14 wt.% Ni-14 wt.% Mm alloy referred to in Sect. 5.3 and Table 5.2. This exhibits an elongation to fracture exceeding 500% at strain rate 7×10^{-1} s^{-1} and temperature

600 °C ($>0.9\,T_m$). This relatively high optimum strain rate and temperature for superplastic deformation appears to be associated with the ultrafine grain size of this material stabilized by nanophase equilibrium intermetallics. Figure 5.17 [5.31] shows that the optimum strain rate for superplasticity evidently does increase very markedly with decrease in grain size of the material.

References

5.1 J.J. Gilman: J. Appl. Phys. **46**, 1625–1633 (1975)
5.2 (a) L.A. Davis: In *Rapidly Quenched Metals*, ed. by N.J. Grant, B.C. Giessen (MIT Press, Cambridge, MA 1976) pp. 369–391
(b) L.A. Davis: In *Glass – Current Issues*, ed. by A.F. Wright, J. Dupuy (Nijhoff, Amsterdam 1985) pp. 94–124
5.3 (a) J.C.M. Li: In *Treatise on Material Science and Technology*, ed. by H. Herman (Academic, New York 1986)
(b) J.C.M. Li: In *Chemistry and Physics of Rapidly Solidified Materials*, ed. by B.J. Berkowitz and R.O. Scattergood (TMS, Warrendale, PA 1983) pp. 173–196
(c) J.C.M. Li: In *Rapidly Solidified Alloys*, ed. by H.H. Liebermann (Dekker, New York 1993) pp. 379–430
5.4 P.R. Rao, V.M. Radhakrishnan: In *Metallic Glasses*, ed. by T.R. Anantharaman (Trans. Tech., Aedermansdorf 1984) pp. 225–248
5.5 A.I. Taub: In *Rapidly Quenched Metals*, ed. by S. Steeb, H. Warlimont (Elsevier, Amsterdam 1985) pp. 1611–1618
5.6 S.K. Das, F.H. Froes: In *Rapidly Solidified Alloys*, ed. by H.H. Liebermann (Dekker, New York 1993) pp. 339–377
5.7 L.A. Davis, S.K. Das, J.C.M. Li, M.S. Zedalis: Int'l J. Rapid Solidication **8**, 73–131 (1994)
5.8 M. Dutoit, H.S. Chen: Appl. Phys. Lett. **23**, 357–358 (1973)
5.9 L.A. Davis, S. Kavesh: J. Mater. Sci. **10**, 453–459 (1975)
5.10 Y.H. Kim, A. Inoue, T. Masumoto: Mater. Trans. JIM **31**, 747–749 (1990)
5.11 H. Nagahama, K. Ohtera, K. Higashi, A. Inoue, T. Masumoto: Philos. Mag. Lett. **67**, 225–230 (1993)
5.12 A.I. Taub, M.R. Jackson, S.C. Huang, E.L. Hall: In *Rapidly Solidified Metastable Materials*, ed. by B.H. Kear, B.C. Giessen (North-Holland, Amsterdam 1984) pp. 389–394
5.13 H. Jones: Proc. 6th Int'l Symp. on Plasticity of Metals and Alloys, Prague (1994). Key Eng. Mater. **97/98**, 1–12 (1995)
5.14 P. Gilman: Met. Mater. **6**, 504–507 (1990)
5.15 A.I. Taub, S.C. Huang, K.M. Chang: Met. Trans. A **15**, 399–402 (1984)
5.16 D.J. Skinner: In *Dispersion Strengthened Aluminium Alloys*, ed. by Y.-W. Kim, W.M. Griffith, (TMS Warrendale, PA, 1988) pp. 181–197
5.17 W.S. Cebulak, E.W. Johnson, H. Markus: Met Eng Quart. **16**(4), 37–44 (1976)
5.18 D.G. Morris: In *Rapidly Quenched Metals*, ed. by S. Steeb, H. Warlimont (North-Holland, Amsterdam, 1985) pp. 1775–1778
5.19 D. Raybould: The Carbide and Tool J. **16**(6), 27–30 (1984)
5.20 Anon: Met. Powder Rep. **46**(11), 3 (1991)
5.21 N. Amano, Y. Odani, Y. Takeda, K. Akechi: Met. Powder Rep. **44**(3), 186–190 (1989)
5.22 T. Hayashi, Y. Takeda, K. Akechi, T. Fujiwara: Met. Powder Rep. **46**(2), 23–29 (1991)
5.23 J. Duszczyk, J.L. Estrada, B.M. Korevaar, T.L.J. de Haan, D. Bialo, A.G. Leatham, A.J.W. Ogilvy: Proc. P/M Aerospace Materials Conference (MPR Publ. Services, Shrewsbury 1987) Paper 26
5.24 A. Inoue: Mater. Trans. JIM **36**, 866–875 (1995)

5.25 T. Masumoto, A. Inoue, H. Yamamoto, J. Nagahora, T. Shibata: Europ. Pat. Appl. 0513654A1 (19th Nov 1992)
5.26 T. Masumoto, A. Inoue, N. Nishiyama, H. Horimura, T. Shibata: Europ. Pat. Appl. 0517094A2 (9th Dec 1992)
5.27 R.S. Mishra, H. Jones, G.W. Greenwood: Int'l J. Rapid Solidification **5**, 149–162 (1990)
5.28 O.A. Ruano, L.E. Eiselstein, O.D. Sherby: Met. Trans. A **13A**, 1785–92 (1982)
5.29 J.K. Solberg, J. Tørklep, Ø. Bauger, H. Gjestlund: Mater. Sci. Eng. A **A134**, 1201–1203 (1991)
5.30 K. Higashi: Mater. Sci. Forum **113/115**, 231–236 (1993)
5.31 K. Higashi: In *Aspects of High Temperature Deformation and Fracture* (The Jap. Inst. of Metals, Sendai 1993) pp. 447–454

6 Magnetic and Electronic Properties of Rapidly Quenched Materials

R.C. O'Handley and H.H. Liebermann

Magnetic materials derive their usefulness from the fact that the magnetization, $\mathbf{M}_s = N\mu_m/V$ i.e. the volume density of atomic magnetic moments, μ_m, can be changed by application of a magnetic field. As temperature increases it becomes increasingly more difficult for an applied field to orient the atomic moments. The net spontaneous magnetization, vanishes at the Curie temperature, T_C (Fig. 6.1). Above T_C, atomic magnetic moments may still exist but their long-range orientations are no longer correlated. Of course with a strong enough field, one could overcome the effects of $k_B T$ and fully align these paramagnetic moments, but at room temperature such fields do not exist. Below T_C, the ability to change the orientation of the saturation magnetization by application of a magnetic field is limited by the *magnetic anisotropy* (the preference for **M** to lie in a particular direction dictated by crystallography, shape and strain). The $M(H)$ curves shown indicate nothing about magnetic anisotropy as they stand; magnetic anisotropy implies a difference in the field needed to magnetize a sample in *different directions*. If significant anisotropy were present, saturation in a *hard direction* would require a larger field than saturation in an easy direction. *Magnetostriction* is the strain (typically measured in parts per million) that accompanies a change in magnetization. The inverse effect, piezomagnetism, a change in the direction of preferred magnetization induced by a stress, is also very important.

Samples of finite size generally break into *magnetic domains*, regions in which the magnetization is saturated, generally along an *easy axis*. Thus a sample can have zero net magnetization even though the atomic moments are still aligned within domains. Domains are separated by *domain walls*, surfaces across which the direction of magnetization changes significantly. Domain walls typically have widths of 10 to 300 nm for hard and soft magnetic materials, respectively. The ease or difficulty with which the flux density, $B = \mu_0(H + M)$, of a sample can be changed is measured by the *permeability* $\mu = \mu_r\mu_0 = d\mathbf{B}/d\mathbf{H}$, and the *coercivity*, H_c, respectively (Fig. 6.2). When domains are present, the net magnetization can be changed without rotating the domain magnetization, $\mathbf{M}_d$, if the domain walls can be moved. Domain wall motion is a strong function of microstructural features and defects having dimensions comparable to the wall width.

The properties of magnetic materials can be grouped into two categories, *fundamental* and *technical* (see Fig. 6.3). Properties such as the saturation mag-

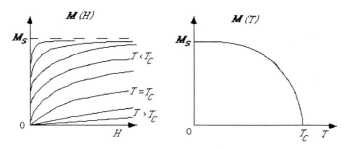

Fig. 6.1. Fundamental properties. Variation of magnetization with field for various temperatures, *left*, and variation of the saturation magnetization with temperature, *right*

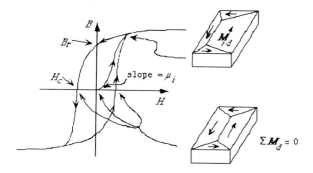

Fig. 6.2. Technical properties. The concept of magnetic domains is illustrated at the *right* while their relation to the magnetization process and other technical properties are illustrated on the *B-H* loop

netization, M_s, the Curie temperature, T_C, the magnetocrystalline anisotropy, K, and the magnetostriction constant, λ_s, are considered to be fundamental. These properties are determined mainly by the electronic structure of the material which in turn is a function of the *short-range order*, i.e. the number, type (chemistry), distance and symmetry of the nearest neighbors about a given atom. Other properties such as those associated with the *B-H* loop, e.g. the permeability, μ_r, and the coercivity, H_c, are of more technical importance. The technical properties are strongly dependent on microstructure (grain boundaries, texture, defects etc.) which in turn is predominantly a result of processing. Fundamental properties are mainly a reflection of the thermodynamics of the material; technical properties reflect more the kinetics of the processing route by which the particular sample was made. Thus one needs a reasonably good grasp of physics, chemistry and materials science to fully appreciate magnetism in rapidly quenched materials.

Fundamental properties of magnetic materials are generally discussed in terms of $\mathbf{M}(H, T)$ while technical properties are generally described by the features of the hysteresis loop, $\mathbf{B}(H) = \mu_0(\mathbf{H} + \mathbf{M})$.

Note that we have discussed magnetism over three different length scales:

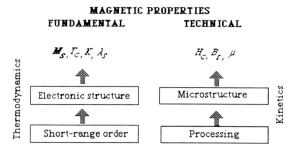

Fig. 6.3. Classification of magnetic properties as *fundamental* and *technical* and the factors controlling those properties. The behavior of the fundamental and technical properties are shown schematically in Figs. 6.1, 2, respectively

atomic magnetic moments, μ_m, domain magnetization, $\mathbf{M}_d$, (typically measured over tens of microns), and macroscopic magnetization, $\mathbf{M}(H)$. Each one is the vector sum of the magnetizations at the next lower length scale.

Nanocrystalline and *amorphous* materials differ significantly from each other and from crystalline materials both in their microstructures and in their atomic structures. Because both microstructure and atomic/electronic structure are important in determining magnetic properties, we must treat these two classes of materials separately. We will consider these materials in terms of the length scale over which atomic order exists and in terms of the length scale of their fluctuations in structure or microstructure. *Quasicrystals* present a special case in which the local atomic structure is neither amorphous nor crystalline (in the classical sense) but is rather quasi-periodic, a mixture of two incommensurate repeat lengths or lattice constants.

Many good reviews of amorphous magnetic alloys are available, covering a variety of physical properties. Some may be found in the bibliography at the end of this chapter.

We will describe some of the important fundamental and technical properties observed in amorphous and nanocrystalline magnetic materials and point out the differences and similarities with crystalline materials.

6.1 Rapidly Quenched Alloys

6.1.1 Amorphous Alloys

Although glass science is quite old, the history of amorphous alloys is comparatively young. Various researchers appear to have stumbled across non-crystalline metallic alloys in their quest for new materials. Examples include the early observations on Ni-P electro-deposits (1930) and superconducting films (1952). However, it is generally accepted that until Pol Duwez began his extensive research on metastable and amorphous alloys at CalTech in the late 1950's, the intrinsic scientific interest and the technological potential of such materials were not widely appreciated.

A ferromagnetic amorphous CoP alloy was first reported in 1965 [6.1] and splat-quenched ferromagnets with attractive soft ferromagnetic properties were reported by Duwez's group in 1966 [6.2, 3]. The subsequent growth of interest in metallic glasses over the next 15 to 20 years was exponential.

The fabrication of amorphous metals is described elsewhere in this book. Essentially all of the successful methods remove heat from the molten alloy at a rate that is fast enough to preclude crystallization of the melt. Clearly some compositions are more viscous and resistant to crystallization than others.

Amorphous alloys of magnetic interest are based either on 3-d transition metals (T) or on rare-earth metals (R). In the first case, the alloy can be stabilized in the amorphous state with the use of glass-forming elements such as boron, phosphorus and silicon:

$T_{1-x}M_x$, with $15 < x < 30$ at.%, approximately.

Examples include $Fe_{80}B_{20}$, $Fe_{40}Ni_{40}P_{14}B_6$, and $Co_{74}Fe_5B_{18}Si_3$.

The late transition metals (TL = Fe, Co, Ni) can be stabilized in the amorphous state by alloying with early transition metals of 4-d or 5-d type (Zr, Nb, Hf):

$TE_{1-x}TL_x$, with x approximately in the range 5 to 15 at.%.

Examples include $Co_{90}Zr_{10}$, $Fe_{84}Nb_{12}B_4$, and $Co_{82}Nb_{14}B_4$.

Rare-earth metals can be stabilized by alloying with transition metals and metalloids:

$R_{1-x-y}T_xM_y$ with x in the approximate range 10 to 25 at.% and y from 0 to 10 at.%.

Examples include GdCo and TbFe (B).

The 3-d transition metal based alloys are generally soft magnetic materials while the rare-earth based alloys can be tailored to span a range from hard (permanent) magnets to semi-hard materials suitable for use as magnetic recording media.

6.1.2 Nanocrystalline Alloys

Nanocrystalline alloys are comprised primarily of crystalline grains having at least one dimension on the order of a few nanometers. Such an extremely small crystallite size can result in novel and/or enhanced magnetic properties.

Nanocrystalline alloys can be fabricated by quenching certain alloy compositions from the melt at a rate insufficient to achieve a non-crystalline structure. Alternatively, they may be made by heat-treating an amorphous alloy precursor Amorphous alloys inherently have a uniform distribution of constituent elements, a condition suitable for the formation of nanocrystalline alloys.

The majority volume fraction of nanocrystalline alloys is a random distribution of crystallites having dimension(s) of order 5 to 40 nm. That volume fraction which is not crystalline is typically situated as a grain boundary phase in which the nanocrystalline grains are embedded. This amorphous phase has a chemistry different from that of the parent amorphous alloy.

Isolated magnetic nanocrystalline particles will contain no domain walls if it costs more energy to create a domain wall than it costs to support the magnetostatic energy of the single-domain state. Single-domain particles can show a broad range of coercivities from very large values of order $2K_u/M_s$ (close to the upper limit for single-domain particles) to essentially zero when the particles become so small that thermal energy is sufficient to flip the magnetization direction ($k_B T \geqslant K_u V$). Such thermally induced lossless magnetic behavior is called superparamagnetism. When nanocrystalline particles are part of a composite, their behavior depends not only on particle size and properties but also on the properties of the intergranular phase. If the intergranular material is non-magnetic, the nanocrystals behave essentially like single-domain particles, interacting only by their weak magnetic dipole fields. In this case they can exhibit either hard or soft magnetism, depending on the size and anisotropy of the particles. One example is Fe–Nd–B alloys heat treated to form nanocrystallites of the high-anisotropy $Fe_{14}Nd_2B$ phase separated by a nonmagnetic B and Nd enriched phase, (Sect. 6.4.2). If the intergranular phase is magnetic, the behavior can be much more complicated but is generally magnetically soft. The reason for this is that, due to the fine grain size of nanocrystalline alloys, the magnetic exchange interaction can overwhelm the crystalline anisotropy in each nanocrystallite. Thus the direction of magnetization can be continuous from grain to grain. The result is a net low magnetic anisotropy of nanocrystalline alloys just as it is in amorphous alloys.

A widely-studied nanocrystal-forming amorphous alloy has nominal chemistry $Fe_{73.5}Cu_1Nb_3Si_{13.5}B_9$. Metallurgically, the effectiveness of developing a nanocrystalline phase is established by the Cu inducing massive nucleation and the Nb slowing grain growth during transformation from the amorphous to the nanocrystalline alloys. The nanocrystalline phase in these alloys is α-Fe_3Si which occupies some 70–80 vol. %. The grain boundary phase, which is amorphous, has a thickness of about 1 nm.

6.2 Fundamental Magnetic Properties

6.2.1 Magnetic Moments and Curie Temperatures

Figure 6.4 below shows the variation of saturation moment per Transition Metal (TM) atom (4.2 K) as a function of TM content for amorphous alloys based on boron, $TM_{80}B_{20}$, and on phosphorus, $TM_{80}P_{20}$. The variation of magnetic moment in crystalline alloys is shown as a dotted line for reference.

Reasonably large magnetic moments can be realized in a variety of amorphous alloys based on iron, cobalt and/or nickel. The reduced moments of

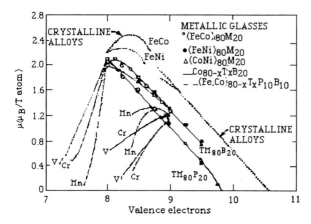

Fig. 6.4. Variation of magnetic moment per transition metal atom in crystalline and amorphous alloys as a function of number of valence electrons. Valence electron concentrations of 8, 9 and 10 correspond to Fe, Co (or $Fe_{0.5}Ni_{0.5}$) and Ni, respectively. The data for crystalline materials are referred to as the Slater-Pauling curve [6.4]

amorphous alloys compared to crystalline alloys reflect the presence of the metalloid (M) atoms, B, P, Si, etc., which are needed to stabilize the glassy state (see Sect. 7).

The saturation moment and Curie temperature vary as the TM/M ratio deviates from 80/20. Figures 6.5 and 6.6 show the variation of saturation magnetic moment per TM atom and Curie temperature with metal/metalloid ratio in amorphous Fe-based and Co-based systems. In both cases the moments increase with decreasing metalloid content, extrapolating to values close to those of bcc Fe (2.2 μ_B/atom) and hcp Co (1.7 μ_B/atom), respectively.

The Curie temperatures of most cobalt rich metallic glasses exceed their crystallization temperatures so their trend with metalloid content is shown over a limited metalloid range (Fig. 6.6). T_C for cobalt rich glasses increases for decreasing cobalt content much as does the magnetic moment. The behavior of the Curie temperature for the Fe-based glasses is quite different (Fig. 6.5): T_C decreases with decreasing metalloid content for both FeB and FeP glasses.

The overall trend in T_C for the amorphous Fe–B alloys follows the trend of the T_Cs for the crystalline alloys, Fe_3B, Fe_2B and FeB (Fig. 6.7). Note that the extrapolation of the Curie temperature for Fe–B with decreasing boron content goes toward values typical of densely packed fcc Fe. This is presumably because the local coordination of an amorphous alloy (12 nearest neighbors) is more like that of an fcc crystalline structure than it is like that of a bcc crystalline structure.

Changes in magnetization and Curie temperature with M-atom type are generally weaker than changes with TM/M ratio or with TM makeup.

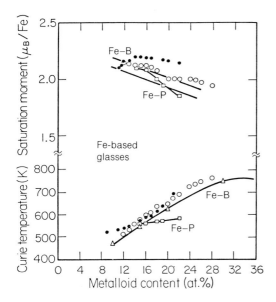

Fig. 6.5. Variation of magnetic moment and Curie temperature with metal-metalloid ratio in binary Fe–B and Fe–P metallic glasses [6.4]

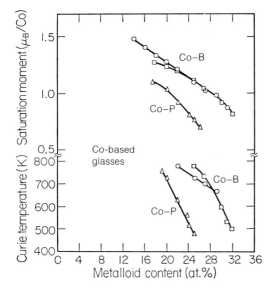

Fig. 6.6. Variation of magnetic moment and Curie temperature with metal-metalloid ratio in binary Co–B and Co–P metallic glasses [6.4]

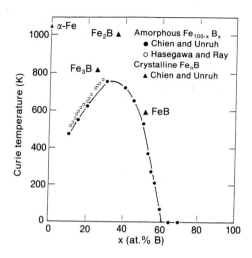

Fig. 6.7. Curie temperatures of amorphous Fe–B alloys vs. boron concentration and for relevant crystalline Fe–B phases [6.5–7]

6.2.2 Magnetic Anisotropy

Crystalline magnetic materials are more easily magnetized in certain crystallographic directions than in others. For example, the $\langle 100 \rangle$ directions are easy magnetization directions in bcc Fe while the $\langle 111 \rangle$ directions are easy for fcc Ni. Anisotropy in the magnetization process implies that the energy of a magnetic material depends upon the direction of the magnetization vector with respect to some preferred direction in the sample. Thus the sample energy density may be expanded in terms of spherical harmonics:

$$E_K = A_0 + A_l Y_l^m(\theta, \phi) \ldots,$$

which for a uniaxial material has the form

$$= A_0 + A_l \alpha_z^2 \ldots,$$

or in a more familiar form

$$= K_0 + K_1 \sin^2\theta + \ldots, \qquad (6.1)$$

where K_0 is the energy expended to saturate the magnetization in any direction; K_1 is the additional energy cost of saturating the magnetization in a direction $\theta = 90°$ away from an easy axis, $\theta = 0°$. Clearly, magnetic anisotropy of crystalline origin is not a factor in amorphous alloys. However, the ease or difficulty or reaching saturation in a given direction is also affected by sample shape, by strain-induced anisotropy, or by field-induced atomic ordering [vide infra]. These factors still operate in non-crystalline materials and they are described by the appropriate higher order terms in (6.1). A uniaxial shape or strain gives rise to a uniaxial magnetic anisotropy.

In nanocrystalline alloys, each grain has preferred directions of magnetization based on its crystallography, shape and state of stress. The random distribution of these easy axes over an ensemble of grains may lead to a near-zero anisotropy, on average, under certain conditions. We will treat this more complicated situation later.

6.2.3 Magnetostriction

The most significant contribution to the macroscopic anisotropy in $3d$-base metallic glasses, after sample shape, is due to internal stresses. Stress or strain affects the magnetic anisotropy of a material to an extent governed by the magnetostriction coefficient, λ_s, or by the magnetoelastic coupling coefficient, B. That is, the energy depends not only on the direction of magnetization relative to an axis in the sample [as in (6.1)] but also on its direction relative to the axes of a strain, e_{zz},

$$E_K = A_0 + A_1 \alpha_z^2 \ldots + B e_{zz} \alpha_z^2 \ldots \qquad (6.2)$$

The magnetoelastic coupling coefficient B is a stress of magnetic origin; it describes the strain-dependence of the magnetic anisotropy energy density. B is proportional to the magnetostriction coefficient, the amount a material strains as the magnetization direction is rotated:

$$\lambda_i = -B_i/C_{ij} \qquad (6.3)$$

where C_{ij} is a combination of elastic constants of the material.

Because the magnetocrystalline anisotropy of amorphous alloys is zero, their soft magnetic properties are controlled largely by stress-induced anisotropy and hence scale with the magnetostriction coefficient. Figure 6.8 below,

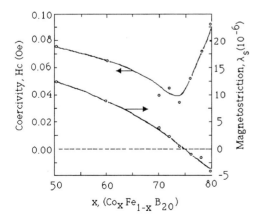

Fig. 6.8. Coercivity and magnetostriction vs composition in amorphous Co-rich alloys

shows how the coercivity in amorphous Co–Fe–B alloys reaches a minimum near the composition for which λ_s zero [6.9].

Such zero-magnetostriction alloys are important for some of the same reasons that make the crystalline permalloy composition, $Ni_{81}Fe_{19}$, important, namely low coercivity, high permeability, and ease of magnetization.

The ternary diagram in Fig. 6.9 shows the compositional variation of the magnetostriction constant over a field of Fe–Co–Ni-base amorphous alloys. What we see is that the magnetostriction is of order 30×10^{-6} for iron rich glasses and drops to zero with cobalt additions near an Fe/Co ratio of approximately 4/76. This ratio is close to the Fe/Co ratio for zero magnetostriction in crystalline alloys. Zero magnetostriction ternary compositions are found along the solid line. The data in this figure are not strongly dependent on the type of metalloid used in the alloy.

There are also important $\lambda_s = 0$ compositions in the (FeCoNi)Zr system (no metalloids).

Because of the technological importance of low magnetostriction, extensive efforts have been made to reduce the magnetostriction of amorphous alloys based on iron. Iron has a higher saturation magnetization and is relatively abundant in nature compared to cobalt; the magnetostriction of $Fe_{80}B_{20}$ is approximately 32×10^{-6} at ambient temperature. Room temperature magnetostriction of $(FeNi)_{80}B_{20}$ glasses was shown to scale with M^2 so that $\lambda_s = 0$ could only be approached with a loss of magnetization. Nevertheless, it has been found that with substitutions of Cr or Nb for Fe, it is possible to achieve $\lambda_s < 5 \times 10^{-6}$ while retaining reasonable values of magnetization. In these Fe–TE–metalloid alloys the decreased room temperature magnetostriction is due partially to the suppressed Curie temperature [6.10].

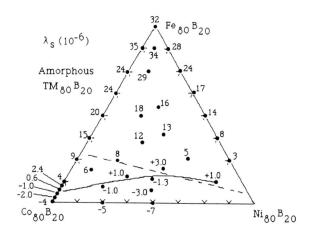

Fig. 6.9. Saturation magnetostriction at room temperature for amorphous alloys $(Fe-Co-Ni)_{80}B_{20}$. The *solid line* shows course of zero magnetostriction compositions and the *dashed line* shows predictions based on the split band model [6.8]

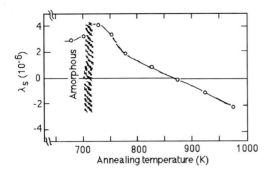

Fig. 6.10. Magnetostriction of FeTaC in amorphous phase *left*, and through increasing stages of crystallization, *right* [6.11]

In nanocrystalline alloys, the sum of the volume-weighted magnetostrictions of the nanocrystalline-grain phase and of the amorphous-grain-boundary phase can result in a net zero alloy magnetostriction even when the starting amorphous alloy shows strong magnetostriction. Figure 6.10 shows the evolution of magnetostriction from amorphous $Fe_{81.4}Ta_{8.3}C_{10.3}$ ($\lambda_s \approx +2.5 \times 10^{-6}$) through various annealing temperatures to $\lambda_s \approx 0$ for 20 min at 870 °C [6.9]. The positive magnetostriction of the residual amorphous phase is balanced by the negative magnetostriction of the α-Fe nanocrystalline phase. In the FeCuNbBSi nanocrystalline system, the α-Fe_3Si nanocrystals have $\lambda_s < 0$ [6.12]. This absence of magnetostriction makes such nanocrystalline alloys attractive in numerous soft magnetic applications including high frequency electronic components. Along with the approach to zero magnetostriction, the amorphous-to-nanocrystalline transformation also results in increased magnetic permeability, further making nanocrystalline alloys attractive in soft magnetic applications. The softest magnetic properties are obtained for the smallest nanocrystalline grain sizes.

6.3 Domains and Technical Properties of Amorphous Alloys

6.3.1 Domains

This section deals exclusively with amorphous alloys for two reasons. First, nanocrystalline alloys are generally comprised of single-domain particles and hence their domain structure is tied to the grain structure. Second, because of the small size of the domains in nanocrystalline materials they are difficult to image magnetically. Figure 6.11 compares the domain structure of polycrystalline permalloy with that of an amorphous iron-rich alloy [6.13]. The SEM

Magnetization direction

Fig. 6.11. Magnetic domains in polycrystalline permalloy foil, *above*, contrasted with those in an amorphous FeBSi alloy, *below*. Permalloy images taken by Type II contrast (45 degree tilt of sample stage relative to e-beam) in SEM; the grain size is 30 μm. The amorphous alloy was imaged by scanning electron microscopy with spin polarization analysis (SEMPA) [6.13]; the field of view here is 1 mm

image of the polycrystalline $Ni_{81}Fe_{19}$ sample shows several domains with rectilinear domain walls in one grain. (The magnetization directions in the other grains are not at the correct angle to show magnetic contrast in this titled-specimen, Type II contrast, image). The domains in the amorphous sample were imaged using Scanning Electron Microscopy with spin Polarization Analysis (SEMPA) [6.14]. Note the absence of rectilinear domain walls in the non-crystalline material. This reflects the absence of long-range crystalline order in metallic glasses and consequently the absence of magnetocrystalline anisotropy. The magnetization tends to follow the local easy axis which, in this case, is largely dictated by internal stress.

The walls separating domains move in response to a field and are responsible for most of the technical properties of magnetic materials. The ease or difficulty with which domain walls move is governed by defects and other inhomogeneities in the material. For a defect to strongly impede wall motion, it should have magnetic properties very different from those of the matrix and it should have dimensions comparable to the domain wall width. Amorphous alloys are characterized by very broad domain walls.

6.3.2 Coercivity

The coercivity of amorphous materials is due to variations in domain wall energy, σ_{dw}, with position in the material. Application of a magnetic field exerts a pressure on a 180° domain wall equal to $2M_s H \cos\theta$, where θ is the angle between the field and the magnetization direction. If the pressure is such as to move the wall to a position of larger wall energy density, the field pressure will be opposed by the wall energy density gradient, $d\sigma_{dw}/dx$. The greater the fluctuations in wall energy with position, the higher the field needed to move the wall: H_c is proportional to $(2M_s)^{-1} d\sigma_{dw}/dx$. For very small defects, the wall energy gradient varies linearly with defect size divided by the domain wall width [6.15]. For amorphous alloys the domain wall width ranges approximately from 0.2 to 2.0 μm.

For the most part, amorphous alloys are homogeneous, i.e. there are no grains, no grain boundaries and no precipitates of any appreciable size. Because these alloys are rapidly quenched from the melt, most impurities tend to remain in solution rather than precipitating out. Thus chemical or structural inhomogeneities (except for surface roughness, pin holes and strain fields) have a scale less than 2 or 3 nm. Because the domain walls are wide in an amorphous alloy and the defects are narrow, there is little pinning of domain walls on defects in amorphous materials and H_c can be very small, the permeability can be very large. Transition metal-based amorphous alloys make excellent soft magnetic materials.

The atomic disorder of amorphous alloys does increase the electrical resistivity to values of order $120\,\mu\Omega\cdot cm$ because conduction electrons probe the material on a scale of Ångstroms (domain walls interact with features in the

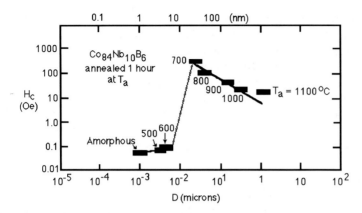

Fig. 6.12. Variation of coercivity of amorphous CoNbB with annealing temperature and with grain size, D, in the crystallized phase. The grain size was measured by TEM [6.16]

material measuring thousands of Ångstroms). Higher electrical resistivity suppresses eddy currents which tend to be induced when rapid flux changes occur. This makes amorphous alloys attractive for higher frequency operation.

The coercivity of nanocrystalline alloys can vary greatly, depending on the magneto-crystalline anisotropy of the grains, on the grain size and on the nature of the grain boundary phase. Figure 6.12 shows a classic example of the variation of H_c over nearly four orders of magnitude upon devitrification of an amorphous cobalt-niobium-boron amorphous alloy. This behavior will be understood when we treat the random anisotropy model below.

6.3.3 Magnetic Hardening

It is interesting that many of the physical mechanisms responsible for metallurgical hardening are also operative in magnetic "hardening", that is, the increase of coercive field on mechanical working of a magnetic alloy. For example, the presence of solid solution strengthening elements, second phase particles, and dislocations or antiphase boundaries in alloys all contribute to both increased mechanical and magnetic hardening. On the other hand, amorphous metallic alloys present a counterexample to this association between mechanical and magnetic hardening. That is, amorphous metallic alloys are very hard mechanically (HV ≈ 1000) while being very soft magnetically ($H_c \leqslant 0.1$ Oe is typical). Ultimately, magnetic hardening can result in permanent magnetism. The utility of permanent magnets is largely determined by their "energy product," $(BH)_{max}$. The maximum energy product is related to the area inside the second quadrant of the B-H loop. As such it is proportional to the product of the coercive field and remanence.

Magnetic hardening stems from the pinning of magnetic domain walls. This can be achieved by any of the five physical mechanisms described below.

(1) intrinsic fluctuations of exchange energies and local anisotropies can result in less uniformly-defined direction of magnetization. This would increase the likelihood of magnetic domain wall pinning.
(2) chemical short range order, resulting from localized clusters of atoms, can also cause an orientational distribution of local easy axes. As with intrinsic fluctuations, chemical short range order can result in magnetic domain wall pinning and an associated lack of an overall definition of magnetization direction.
(3) atomic rearrangements can result in relaxation effects, which can affect domain wall pinning and their motion.
(4) surface features such as scratches, pits, etc. act as magnetostatic pinning centers for magnetic domain walls.
(5) defects within alloys having high magnetostriction can result in volume pinning of magnetic domains.

In actuality, all of these pinning mechanisms operate simultaneously. Most often, though, one or two of these mechanisms dominate over the others.

6.3.4 Induced Anisotropy

It is possible to alter the technical properties of a metallic glass by field annealing. Dramatic micrographs of the magnetic domains in amorphous ribbons annealed in longitudinal and transverse fields illustrate the magnetic consequences of field annealing (Fig. 6.13). What is the mechanism behind this effect?

When a magnetic material is heated below its Curie temperature but at a temperature high enough to allow substantial short-range atomic mobility, the thermal motion of the atoms will result in a slight biasing of the local structure toward its more stable atomic configuration with respect to the local direction of magnetization (Fig. 6.14). If a field is applied during the annealing process, then the magnetization may exhibit a long-range orientational order that will result in local atomic rearrangements that have a long-range orientational correlation with the magnetization direction. Upon cooling, the magnetization of the entire sample will have a tendency to orient in the direction it had during the annealing process. The enhanced atomic mobility in the amorphous state allows for pair-ordering in certain cases at relatively low temperatures compared to crystalline alloys. The role of the metalloids in this process should clearly be important because of their high mobility and strong chemical interaction with the T metals. If they assume a non-random orientational distribution around the T sites, they may favor magnetization in a particular direction.

Annealing in the absence of a field also induces a local magnetic anisotropy along the axis of the local magnetization. Thus, when a domain wall exists during annealing, it will be stabilized in that position because moving it requires rotation of the local moments from their stable orientations. Such wall pinning is illustrated by the M-H loops shown below in Fig. 6.23 on amorphous cobalt-rich alloy ribbons.

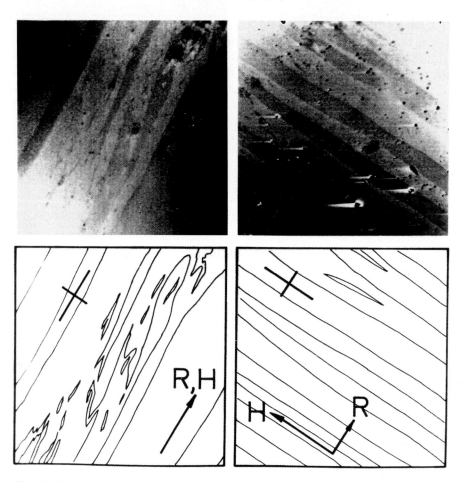

Fig. 6.13. Domain images of a cobalt rich amorphous alloy as cast, *left*, and annealed in a transverse field, *right* [6.17]

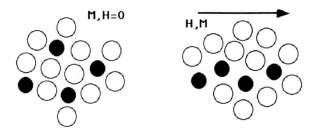

Fig. 6.14. Schematic representation of short range order in a binary alloy. Left, no applied field and no net magnetization. Right, magnetization ordered under influence of field, leads to atomic pair ordering with a directionality correlated to the field direction

In the case of rare-earth containing amorphous films, a uniaxial anisotropy often can be developed by controlling the sputtering conditions during film growth. Strong perpendicular anisotropy, useful in magneto-optic recording materials, can be achieved. Film growth with a columnar morphology is believed to be a source of the perpendicular anisotropy in sputtered Gd–Co and Tb(FeCo) films.

Uniaxial magnetic anisotropy can be induced also by stress-annealing. In contrast to field-induced anisotropy, stress-induced magnetic anisotropy can be induced even above the Curie temperature. This characteristic indicates that the source of stress-induced anisotropy is due to viscoelastic effects rather than to magnetoelastic effects. The magnitude of stress-induced anisotropy can be much greater than that of magnetic field-induced anisotropy. In fact, combined stress-field anisotropy anneals on amorphous alloys have resulted in an induced anisotropy strength greater than either of the methods individually.

6.4 Magnetism and Short-Range Order

6.4.1 Ingredients of Short-Range Order

We now address the issue of the differences and similarities between the fundamental magnetic properties of amorphous and crystalline alloys as depicted primarily in Fig. 6.4. Early interpretations of such Slater-Pauling-like curves of magnetic moment variation with composition (or electron concentration) invoked charge transfer from the valence band of the glass former to the 3d band of the T species to explain what appears to be a shift of the data for the glasses relative to that for pure crystalline alloys. The assumption was that the addition of metalloid p electrons to the 3d band would fill these bands prematurely as 3d valence electron concentration increases. The naive charge transfer models soon gave way to a more realistic understanding of compositional effects in terms of (sp)-d bonding. This (sp)-d bonding or hybridization is evident in theoretical treatments and experimental data on electronic structure and bonding of metallic glasses.

Hund's rules describe the formation of magnetic moments on isolated atoms. Orbitals fill so as to maximize first the sum of their spin quantum numbers. The basis for this rule is found in the Coulomb interactions that lower the energy of electrons whose motions are correlated by occupying orbitals of different angular momentum. Essentially Hund's rule can be described as an intra-atomic interaction that favors parallel spins.

The question before us is "What happens to the intra-atomic interaction that favors the formation of magnetic moments when atoms are put together to form solids?"

It is now well established that fundamental magnetic properties – even in metallic 3d alloys – are predominantly determined by the immediate environment about the atoms. Although many properties of ferromagnets (saturation moment, magnetic anisotropy, ordering temperature) can be calculated in the context of band theory, it is the dependence of the electronic energy levels on *local* structure rather than the Bloch nature of the wave functions that is critical.

Four aspects of short-range order can be identified as important to magnetism:

the *number,*
type,
distance and
symmetry

of the nearest neighbors about a given site. This is illustrated in simple terms by the Stoner criterion for the existence of a local moment:

$$I(E_F)D(E_F) > 1, \tag{6.4}$$

or by the molecular field expression for the Curie temperature:

$$T_C = J(r)Z_T S(S + 1)/3k_B. \tag{6.5}$$

Here $J(r)$ is the *distance*-dependent interatomic exchange integral, Z_T is the coordination *number* (presumably by strongly magnetic species) about the transition metal, T, site, S is the atomic spin quantum number, $I(E_F)$ is the Stoner integral evaluated at E_F reflecting what remains in the solid of intra-atomic (Hund's rule) exchange, and $D(E_F)$ is the electronic density of states at the Fermi energy. The *number, type* and *distance* of nearest neighbors enter (6.4) and (6.5) explicitly through Z_T and $J(r)$ and implicitly through $D(E_F)$ and $I(E_F)$. The *symmetry* of the nearest neighbor arrangement affects $D(E_F)$ and $I(E_F)$ by changing the degeneracy and hence the distribution of the electronic states. Thus local magnetic moment formation is determined by the extent to which intra-atomic exchange is free to operate. Whereas a full moment equal to half the population of the orbital can be achieved in an isolated atom (e.g., $\mu_{Fe^{3+}} = 5\mu_B$ per Fe^{3+}), when the energy levels are broadened by bond formation to the extent that they overlap, the local magnetic moment is reduced (e.g., $\mu_{Fe} = 2.2$ μ_B in bcc Fe). The Curie temperature reflects more cooperative, inter-atomic effects as well as the magnitude of the local moment.

6.4.2 Random Local Anisotropy

Magnetic anisotropy arising from long-range crystallinity is clearly absent in amorphous alloys. However, the same 'crystal-field' or more accurately 'local-field' that gives rise to magneto*crystalline* anisotropy is effective in non-crystalline materials on a scale of a few nanometers and in nanocrystalline materials on the scale of the grain size. The sign and strength of the coupling between this local crystal field and magnetic moment orientation is determined largely by spin-orbit interactions. The degree to which this *local*-anisotropy field affects

6.4 Magnetism and Short-Range Order

macroscopic behavior on the one hand or gets averaged out due to fluctuations in orientation of the local 'easy axes' on the other is important to an appreciation of technical magnetic properties of both amorphous and nanocrystalline materials.

The local crystal field energy is modeled by a uniaxial (dipolar) term of strength D. The orientation of the easy direction of this uniaxial anisotropy fluctuates with a correlation length l determined by the local structure: l is a few nanometers for amorphous alloys and up to several tens of nm for nanocrystalline alloys. We assume the local magnetic moments to be coupled to each other by an exchange interaction of stiffness A expressed by the form $A[\nabla \mathbf{m}(\mathbf{r})]^2$ where $\mathbf{m}(\mathbf{r}) = \mathbf{M}(\mathbf{r})/M_s$ is the local reduced magnetization [6.18, 19].

Given a strength for the local anisotropy, D, it is important to know the orientational correlation length, L, of the local magnetic moments. That is, how closely can the exchange-coupled magnetic moments follow the short-range (l) changes in local easy axis orientation given by the unit vector $\mathbf{n}(\mathbf{r})$. Mathematically the problem reduces minimizing two competing terms in the free energy, F:

$$F = A[\nabla \mathbf{m}(\mathbf{r})]^2 - D\{[\mathbf{m}(\mathbf{r}) \cdot \mathbf{n}(\mathbf{r})]^2 - 1/3\}. \tag{6.6}$$

Clearly the first term can be expressed as A/L^2 provided $L \gg l$. The strength of the random local uniaxial anisotropy expressed by the second term can be evaluated using random walk considerations [6.20]. The *macroscopic* anisotropy can then be expressed as $D(l/L)^{3/2}$, a scaling down of the *local* anisotropy by the ratio $(l/L)^{3/2}$. Energy minimization of F with respect to L then gives

$$L = 16A^2/(9D^2l^3) \approx 10^4/D^2, \tag{6.7}$$

where we have used $A = 10^{-11}$ J/m and $l \approx 2$ nm (amorphous alloys). For a 3d-based amorphous alloy D is of order 5×10^4 J/m^3 while for 4f-rich alloys it is of order 10^6 J/m^3. Hence for these two cases

$L = 20$ μm   3d-based
$L = 20$ nm   4f-based

These results indicate that in 3d-based amorphous alloys exchange stiffness can maintain local moment orientational correlation up to 10 μm despite changes in the local anisotropy direction, whereas in 4f-based amorphous alloys the local magnetic moment may fluctuate over a much shorter range possibly approaching a few nanometers (following the local anisotropy field). Pictures of ferromagnetic domains in 3d-based amorphous alloys (e.g., Fig. 6.11b) support the first estimate and the dispersion of rare-earth moment directions observed in many R–T amorphous alloys supports the second.

For 3d-based nanocrystalline alloys (assuming $l \approx 20$ nm and $D \approx 10^4$ J/m^3) (6.7) gives $L \approx 200$ nm. Thus in 3d-based nanocrystalline systems, the magnetization varies over a length scale about ten times the assumed grain size, the local anisotropy is averaged out and the alloy is soft. Increasing the anisotropy by even a factor of 3 (over the assumed 10^4 J/m^3) lowers L enough that the magnetization is affected by the anisotropy in each grain and H_c increases.

The technical implications of these estimates of L lie in the wide range of macroscopic magnetic anisotropies and coercivities that can be realized in amorphous and nanocrystalline alloys (Fig. 6.12). When $L \gg l$, the effective macroscopic anisotropy is small and soft magnetism is observed. When L/l decreases toward unity, the magnetization vector is more strongly constrained by the orientation of the local anisotropy.

6.5 Electronic Structure of Amorphous Alloys

Under equilibrium processing conditions, local atomic arrangements are essentially determined by thermodynamics which is governed by the nature of interatomic interactions (e.g., chemical ordering) and entropy. Non-equilibrium conditions include rapid heat removal, rapid solidification front velocity, or solid-state treatment in a time-temperature regime where the time needed for diffusion to the equilibrium crystalline state exceeds that required to reach metastable states. A non-equilibrium fabrication process can override the tendency for the atoms to order locally the way they do in the equilibrium crystalline ground state. In such cases entropy plays a more significant role in the local structure obtained than it does in the equilibrium phase. Materials having new structures (e.g., metastable crystalline materials or metallic glasses) may result, depending upon the alloy thermodynamics and on the processing kinetics. Their constrained, non-equilibrium local atomic arrangements determine the electronic structure which, in turn, governs the fundamental physical properties.

Amorphous materials are not random in their atomic structure. They lack long-range order but short-range atomic order exists, just as it does in the liquid state. Because the compositions of amorphous alloys are often close to eutectic compositions, their short-range order is characterized by a competition between the two bracketing stable phases which are responsible for the existence of the eutectic. It is this competition that inhibits crystallization to one of the two adjacent stable crystalline phases. Hence the electronic structure of an amorphous alloy on a local scale (Fig. 6.15, left, center) need not resemble that of any related crystalline composition. Further, because of the long-range orientational disorder in the local structural units, the features of the electronic structure that may exist on a local scale are angle-averaged on a macroscopic scale (Fermi surface, Fig. 6.15, right).

6.5.1 Chemical Bonding

The chemical interactions that affect the electronic states and determine the physical properties of a material can be described by two limiting bonding types: polar bonds and covalent bonds. Covalent bonds are formed between orbitals on two atoms A and C (anion and cation) that have similar elec-

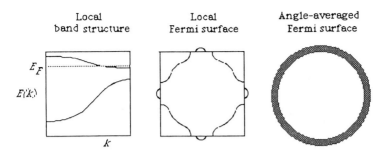

Fig. 6.15. *Left*, schematic representation of electronic structure in a small region (1 nm)^3 of an amorphous alloy and its corresponding Fermi surface, *center*. At *right* we display the effect of adding local Fermi surfaces having random orientation

tronegativities (i.e. similar electronic energies ($E_A = E_C$), as well as satisfying symmetry and overlap conditions. In a covalent bond, charge is delocalized from each of the atomic sites and builds up between the atoms. Bonding and antibonding hybrid orbitals are created. In alloys containing transition metal species, covalent bonds formed between partially occupied valence orbitals (i.e., near E_F) will often involve magnetic states. If d-states are involved they become more delocalized as a result of covalent bonding. This delocalization results in a loss of d-character (hence weaker intra-atomic exchange $I(E_F)$ and in a suppression of $D(E_F)$; both of these effects weaken magnetic moment formation (Sect. 6.4.1).

Polar bonds are formed between orbitals that differ significantly in their electronegativity ($E_A \neq E_C$). The orbitals must also satisfy symmetry and overlap conditions. In the formation of a polar bond, charge is transferred from the orbital of higher energy (lower electronegativity) to that of lower energy. As a result of this charge transfer the charge in the bond is biased toward the more electronegative species, A. If a polar bond is formed in a metal, the conduction electrons will redistribute themselves to screen the bond charge transfer and maintain some degree of local charge neutrality. The screened polar bond still contributes to the chemical stability of the alloy. However it will only affect the magnetic properties if one of the orbitals involved contributes to the magnetism (e.g., a 3d orbital).

6.5.2 Split d Bands and p-d Bonds

The valence bands of an alloy $A_{1-x}B_x$ often split into two resolvable components having different energies when their atomic number difference is greater than or equal to two. The lower energy states are identified primarily with electrons having greater probability density at the site of the more attractive species (i.e., the anion, which for two metals of the same row is the one with higher atomic number).

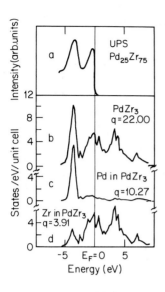

Fig. 6.16. *Top panel*: UPS valence band spectrum of amorphous $Pd_{25}Zr_{75}$. *Lower panels*: calculated total and site decomposed state densities for $PdZr_3$ [6.21, 22]

Figure 6.16 (top panel) shows UPS (Ultra-violet Photoelectron Spectroscopy) data for amorphous PdZr [6.21]. The lower energy (greater binding energy) feature reflects the chemical stabilization due to the more attractive core potential at the Pd site compared with that at the Zr site. The calculated state densities shown in the lower panel are discussed below [6.22].

The connection between split bands and chemical bonding can be seen in UPS spectra on a variety of amorphous TE–TL alloys. The splitting of d-band features in an alloy correlates with the valence difference between the two transition metal species (Fig. 6.17, [6.23]. The deviation of the Cu-based glasses from linear dependence would be remedied by using electronegativity difference as the abscissa instead of valence difference.) The general increase in binding energy with T valence difference correlates with the compound heat of formation as well as with the stability of the glassy phase.

In amorphous T–M alloys (sp)-d bonding dominates. The hybridization between metalloid sp states and metal d states reduces the degree of localization of the d electrons and tends to broaden the d-band into a more free electron-like, less tight-binding-like wave function. This (sp)-d hybridization broadens the d-electron contribution to the DOS in T–M alloys to differing degrees depending upon the extent of hybridization. It also gives rise to the discrete (sp)-d bonding states seen in Fe-based alloys at $-9.5\,\text{eV}$ and lower energies but not seen in α-Fe (Fig. 6.18, [6.24]).

Calculations on finite clusters of atoms allow for the direct study of the effects of local topological order on electronic structure. The self-consistent-field, X-α, scattered-wave, molecular-orbital method has been used to model the electronic structure of tetrahedral clusters of Fe and Ni atoms. Adding central B and P atoms to these clusters reveals the chemical bonding effects important

6.5 Electronic Structure of Amorphous Alloys   175

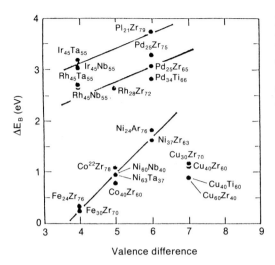

Fig. 6.17. Binding energy shifts ΔE_B of the late T-species d-band maximum upon alloying as a function of the late-early T metal valence difference. The *lines* suggest that 3d, 4d and 5d late T species define three separate alloy groups [6.23]

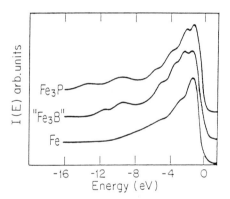

Fig. 6.18. XPS valence-band spectra for the T–M glassy alloys $Fe_3B_{0.9}P_{0.1}$ ("Fe_3B") and Fe_3P compared with that for crystalline Fe [6.24]

in T–M metallic glasses. The formation of metal-metalloid bonding states decreases the amount of d-band character and hence decreases the magnetic moment. This is shown in Fig. 6.19 for Fe_2Ni_2 and Fe_2Ni_2B clusters [6.25]. The occupied (sp)-d hybrid states appear at lower total energy than the corresponding states in the pure metal clusters, accounting for the stabilization upon alloying. Fe–P and Ni–B bonds were found to be stronger than were Fe–B and Ni–P bonds. The p-d bonding states (square brackets at -6, -6.5 eV below the vacuum level, Fig. 6.19) are the small-cluster analogs of the p-d hybrid peak observed by XPS at -9.5 eV (Fig. 6.18).

Electronic states have been calculated for clusters of 1500 atoms arranged as determined by dense random packing [6.26]. The location of spectral features arising mainly from boron s states, from iron-spd bonded with boron p states, and from iron-d states are consistent with those observed by photoemission in Fig. 6.18.

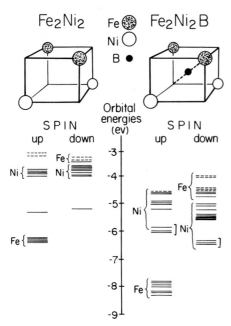

Fig. 6.19. Spin-split electronic states calculated for an Fe$_2$Ni$_2$ tetrahedral clusters without (*left*) and with (*right*) a central boron atom present. *Dashed lines* indicate unoccupied states. States labeled with curly brackets are localized near the species indicated. Unlabeled states near −5.2 eV (*left*) mark the locations of Fe–Ni hybrid states. Square brackets to the *right* of eigenstates mark Ni–B bonding states. Boron atomic 2p levels lie at −3.9 eV [6.25]

Table 6.1 is a summary comparison of the binding energies of the major bonding features in the Fe–B system. Various calculational results are listed for small clusters, large clusters and band structures. Finally, the experimental results are listed. All three calculational approaches, i.e., band structures based on simple close-packed structures, cluster calculations on large random clusters and cluster calculations on small, high-symmetry clusters (Fig. 6.19), agree quite well on the basic chemical physics of bonding and electronic structure. Small clusters always show narrower bands and weaker binding energies for bonding states due to the smaller number of interactions. Also, all the calculations agree well with the experimental XPS data on "Fe$_3$B."

Table 6.1. Binding energy (in eV below E_F) for boron s states and Fe-d-B-(sp) hybrid bonding states as determined by four different calculations and by XPS studies

Method	Boron s states	Fe-d-B-(sp) hybrid	Reference
Fe$_4$B cluster	10.4	7	a
Fe$_2$Ni$_2$B cluster		6.5	b
1500 atom Fe$_{80}$B$_{20}$	12.5	6.8	c
ASW Fe$_3$B band structure		8–10	d
Experimental Fe$_3$B XPS	11.5	9.5	e

a. Collins et al., 1987 [6.27]
b. Messmer, 1981 [6.25]
c. Fujiwara, 1984 [6.26]
d. Moruzzi, unpub.
e. Amamou and Krill, 1980 [6.24]

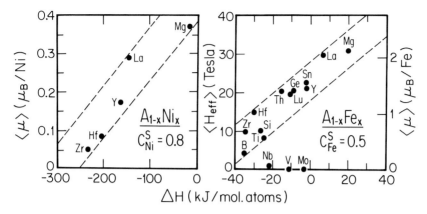

Fig. 6.20. *Left*: Average Ni moment as a function of heat of alloying in various Ni alloys of compositions such that coordination sphere about an average Ni atom is about 80% covered, $CS_{Ni} = 0.8$. *Right*: Average effective hyperfine field or average Fe moment as a function of heat alloying for amorphous Fe-based alloys such that $CS_{Fe} = 0.5$ [6.28]

These calculations and observations on transition metal-metalloid clusters show the mechanisms behind the general observation that chemical bonding weakens magnetism. When magnetic d orbitals are involved in bonding, some states are removed from the d-band to low-lying bonding orbitals that are fully occupied and therefore do not contribute to the magnetic moment. The remaining states in the d-band become delocalized; intra-atomic exchange is less effective in countering the increased kinetic energy of the electrons in a broadened band (6.4) so this lowers the magnetic moment. Figure 6.20 illustrates the manifestations of the competition between bonding and magnetism in systems based on transition metals and non-magnetic metals. The stronger the bonding (the more negative the heat of alloy formation), the smaller are the magnetic moments [6.28].

In summary, these calculations and data flesh out our physical insight into the effects on magnetism of changes in local environment (*number, type, distance,* and *symmetry* of nearest neighbors). They prescribe the principal ingredient favoring magnetism in alloys, whether amorphous, nanocrystalline or crystalline: (1) strongly positive heat of formation to favor like-atom clustering rather than ordering and (2) minimum p-d hybridization (smaller T–M coordination as is usually found for metalloids of small radius).

6.5.3 Electron Transport

Amorphous magnetic alloys have electrical resistivities falling typically in the range from 100 to 150 $\mu\Omega \cdot$ cm (3 times to 5 times those of crystalline magnetic alloys). The loss of structural and chemical order beyond a length of a few

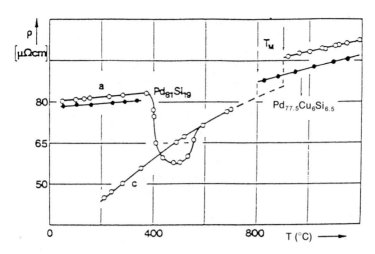

Fig. 6.21. Electrical resistivity of $Pd_{81}Si_{19}$ and $Pd_{77.5}Cu_6Si_{16.5}$ in the glassy, crystalline and liquid states [6.29]

nanometers contributes to electron scattering and limits the electronic mean free path to a distance of order 1 nm. Metallic glasses are characterized by small positive or negative temperature coefficients of resistivity (TCRs). In this sense, metallic glasses bear a close resemblance to liquid metallic alloys of similar composition. Figure 6.21 shows the results for PdCu in the crystallinear amorphous and liquid phases [6.29]. This similarity led to the application of Ziman liquid-metal theory to the understanding of electrical transport in metallic glasses. One immediate outcome of this analysis was a simple explanation of the systematics of occurrence of positive or negative TCRs depending upon the values of the Fermi wave vector relative to the peak of the structure factor (Fig. 6.22, [6.29]). Alloy systems having their Fermi energy near the peak of the structure factor $S(k)$, exhibit lower electrical resistivity at elevated temperatures [less scattering due to reduced peak in $S(k)$] while those with E_F on the wings of $S(k)$ show increased ρ with increasing temperature. This model also correctly describes observations of changes in ρ with structural relaxation in amorphous alloys.

Just as amorphous alloys show low magnetic anisotropy, they also show weak anisotropic magnetoresistance, $(\rho_\parallel - \rho_\perp)/\rho_\perp = \Delta\rho/\rho$ (where $\parallel$ or $\perp$ refers to the field orientation relative to the direction of current). This weak magnetoresistance is due in part to the large value of electrical resistivity that characterizes metallic glasses, $\rho = 100\,\mu\Omega\cdot\text{cm}$.

Metallic glasses do show a strong spontaneous Hall effect, $V_H = R_s(\mathbf{J} \times \mathbf{M})$ where $\mathbf{J}$ is the current density vector and $\mathbf{M}$ the magnetization vector. $\mathbf{R}_s$ is the spontaneous Hall coefficient [6.30]. In a non-magnetic material the application of a magnetic field perpendicular to the primary current can alter the longitudinal resistance $\Delta\rho/\rho$ and induce a Hall voltage in a direction orthogonal to both

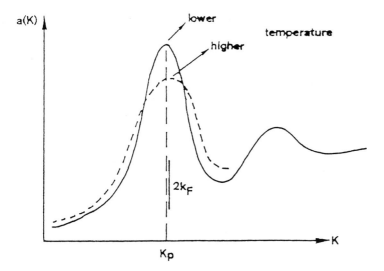

Fig. 6.22. Scattering factor, $a(\mathbf{K})$ vs scattering vector $\mathbf{K}$ for lower and higher temperatures in an amorphous alloy [6.29]

the current and applied field. In non-magnetic materials these anisotropic effects can be understood as consequences of the Lorentz force on the primary current carriers. In a ferromagnetic material these ordinary anisotropic transport effects are generally overshadowed by phenomena with similar geometrical dependences but arising from the much stronger spin-orbit interaction between the current carrier (orbit) and the local magnetization (spin).

The electrical resistivities of nanocrystalline alloys are similar to those of their amorphous precursors. The reasons for this high electrical resistivity are the small grain size (high density of grain boundaries) and the high resistivity of the amorphous, grain boundary phase.

6.6 Applications

Most applications of magnetic materials are based on one or more of four principles: 1) saturated magnetic materials produce fields, 2) unsaturated magnetic materials respond to fields by drawing in field lines, 3) changes in magnetic flux density with time, $\partial B/\partial t$, induces a voltage about the direction of the change, and 4) many physical properties of magnetic materials, e.g., their electrical, magnetic, optical and elastic properties, can be functions of the state of magnetization of the material.

We give brief examples of each type of application then summarize the advantages offered by rapidly solidified materials. 1) Permanent magnets produce magnetic fields whose forces are useful in motors and actuators. 2) The ability of soft magnetic materials to attract fields makes them useful as magnetic shields, magnetic lenses and flux concentrators. 3) The voltage induced by a temporal change in flux density, i.e. Faraday's law, is the basis for operation of transformers and inductors. 4) The dependence of many physical properties of the state of magnetization of a material makes them useful as sensors, transducers, and information storage media. For example, the large permeability of a soft magnetic material drops toward unity as the material is magnetically saturated. The dependence of the optical constants on the state of magnetization gives rise to Kerr and Faraday rotations of the plane of optical polarization. Finally, the Young's modulus, thermal expansion coefficient and even the lattice constants of magnetic materials are appreciable functions of **M**.

It stands to reason then that if the state of magnetization of a material is very stable (a permanent magnet) it may be useful in applications based on principle 1). If the state of magnetization is more easily changed by a field or temperature, the material may be very useful in applications based on principles 2) to 4).

Amorphous metallic alloys lack long-range atomic order and consequently exhibit many characteristics important for a variety of applications. Some of their attractive technical characteristics are listed below.

1) High electrical resistivity for a metallic alloy (100–200 $\mu\Omega \cdot$ cm) due to electron scattering from atomic disorder. High resistivity is important in suppressing eddy currents during high-frequency magnetization reversal.

2) No macroscopic magnetocrystalline anisotropy (residual anisotropies typically amount to $10 \, \text{J/m}^3$ for 3d-based alloys but can approach $10^7 \, \text{J/m}^3$ for certain rare-earth containing alloys). Hence magnetization rotation is relatively easy in the former class of amorphous alloys. Anisotropy fields, H_K, of a few Oersteds are readily achieved.

3) No microstructural discontinuities (grain boundaries or precipitates) on which magnetic domain walls can be pinned. Hence magnetization by wall motion is relatively easy. Coercive fields, H_c, of a few milli-Oersteds are readily achieved.

As a result, ferromagnetic metallic glasses based on 3d transition (T) metals are generally good 'soft' magnetic materials with both low dc hysteresis loss and low eddy current dissipation. In addition, they are characterized by high elastic limits (i.e., they resist plastic deformation) and, for certain compositions, they show good corrosion resistance. Amorphous magnetic alloys containing appreciable fractions of rare earth (R) metals show magnetic anisotropy and magnetostriction that can be varied almost continuously with composition up to very large values. These characteristics combined with the fact that metallic glasses can be economically mass fabricated in thin gauges, has led to broad commercial interest.

Nanocrystalline alloys share the high electrical resistivity of amorphous alloys but offer a much wider range of values of coercivity because of the sensitivity of their average anisotropy, $\langle K \rangle$, to grain size, local anisotropy, and the nature of the grain-boundary phase. Thus while some nanocrystalline alloys show outstanding performance as soft magnetic materials, other nanostructured magnetic alloys make excellent magnetic recording media and still others, high-energy permanent magnets.

Table 6.2 summarizes some of the major applications of rapid solidified magnetic alloys. Selected applications are described in more detail after the table.

Table 6.2. Summary of important technical properties and their underlying physics for some of the major applications of rapidly quenched magnetic alloys

Application	Important properties	Typical compositions
Distribution and power transformers	High $4\pi M_s$ Low magnetic anisotropy High electrical resistivity Low stress sensitivity Good thermal stability	a-$Fe_{78}B_{19}Si_3$
High-frequency transformers	Low magnetic anisotropy High electrical resistivity Zero magnetostriction	a-$Co_{74}Fe_5B_{19}Si_2$ nc-$Fe_{38}Nb_3Cu_2Si_{15}B_3$ nc-FeHfC
Magneto-elastic transducers, EAS sensors	High $4\pi M_s$ Low magnetic anisotropy Strong magnetostriction	a-$Fe_{38}Ni_{38}Mo_2B_{19}Si_3$ a-$Fe_{78}B_{19}Si_3$
Harmonic EAS sensors	Large non-linear permeability Low magnetic anisotropy Zero magnetostriction	a-$Co_{74}Fe_4B_{19}Si_3$
Permanent magnets	High $4\pi M_s$ Large magnetic anisotropy Large coercivity	nc-$Fe_{82}Nd_{12}B_6$ nc-$Fe_{38}Nd_{38}Sm_2B_7$
Magnetic recording heads	High $4\pi M_s$ High permeability Superior wear resistance Good corrosion resistance	a-$Co_{74}Fe_5B_{19}Si_2$ a-$Co_{90}Zr_{10}$ a-$Co_{84}Zr_{12}B_4$
Longitudinal magnetic recording media	High coercivity, controllable squareness, isolated nano-crystalline grains	nc-CoCrTa nc-CoPtCr
Perpendicular magneto-optic recording media	Strong perpendicular anisotropy Moderate $4\pi M_s$ Superior wear resistance Good corrosion resistance	a-$Tb_{24}Fe_{76}$ a-$(Tb_{1-x}Dy_x)_{76}(Fe_{1-y}Co_y)_{24}$

6.6.1 Distribution Transformers

The technological development of amorphous magnetic alloys was motivated largely by their application as cores in distribution transformers, a very high volume market [6.31–33]. Residential power transformers rated at about 25 kVA contain approximately 100 kg of magnetic material. Close to one million distribution transformers are installed annually in the United States alone. Amorphous alloy transformer cores have lower losses and thus save power and reduce the cooling requirements of the transformer construction [6.33]. However, the metastable nature of the amorphous state was a cause for some initial concern for a device which is expected to operate for a minimum of 20 years: what would happen to a transformer if the core crystallized and the losses increased dramatically? The stability of many commercial amorphous magnetic alloys is now known to be sufficient for hundreds of years of operation at 200 °C [6.33] but the question remains a valid one. Installations of amorphous alloy core transformers began in the early 1980s and field performance has been successful; performance has improved with aging. By summer 1990 some 35000 25 kVA distribution transformers containing amorphous alloy cores had been purchased and installed by U.S. utilities.

When the highest permeability is required, as in smaller transformers and other inductive elements, the amorphous cobalt rich alloys represent an attractive alternative to 78% Ni permalloy. One advantage they have over permalloy is their much higher hardness. The high yield strength of amorphous metallic alloys generally makes them much more resistant to plastic deformation and slip-induced anisotropy than crystalline alloys.

Iron-rich amorphous alloys can be used in high frequency applications if they are heat treated to develop a small volume fraction of domain-nucleating, α-Fe precipitates of nanocrystalline dimensions [6.34].

6.6.2 Electronic Article Surveillance Sensors

A growing application of metallic glasses is in the field of Electronic Article Surveillance, EAS, the process of placing remotely detectable tags on items to deter theft. What is needed for EAS is an interrogation zone (usually defined by a magnetic dipole antenna pair) near the exit to an area to be secured. When the magnetic field in the interrogation zone is perturbed by an active tag, the system is alerted. The tags of interest here most often consist of a small strip of amorphous magnetic alloy. Magnetic tags change the characteristics of the field in the interrogation zone in *frequency* or in *time* [6.35].

When a material in the presence of an external ac field shows a response at a different frequency, it is referred to as harmonic generation. Harmonic tags multiply the drive frequency by the non-linear permeability of the tag. Materials for such tags must have very low coercivities and large, nonlinear permeabilities. Cobalt-rich, zero-magnetostriction alloys are most often used for these har-

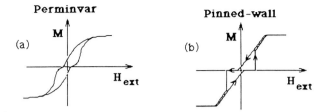

Fig. 6.23. a Typical Perminvar loop. **b** Pinned-wall loop of a small amorphous ribbon [6.35]

monic tags. A novel type of harmonic tag is based on the concept of domain wall pinning. By annealing a strip of soft amorphous alloy in zero field, the domain walls are stabilized in their demagnetized locations [6.36]. After annealing, a small non-zero field is required to free or de-pin the walls. When this field threshold is exceeded, the walls snap to a new position with a resultant sharp change in magnetization. This pinned-wall behavior (Fig. 6.23) is the same effect in a small sample comprised of a few magnetic domains as the well-known perminvar effect in larger multi-domain samples.

Some materials respond to a pulsed excitation field in a way that persists for a time after the excitation field is off. Resonant magnetoelastic tags operate in this way. When a magneto-elastic tag is excited by a primary signal for a period of time, it stores magnetic energy in a coupled magneto-elastic mode. Once the excitation field is turned off, the magnetoelastic tag "rings down" in a characteristic way that allows its signal to be separated in time from the drive signal as well as being distinguished from the signals of most other possible magnetic objects passing through the interrogation zone. The requirement here is for a magnetically soft material having non-zero magnetostriction.

6.6.3 Magnetic Recording Media

Most hard disks for magnetic information storage use a thin film of nanocrystalline CoCrTa or CoPtCr alloy to store the magnetic information. Grain sizes are typically 30 nm and the grain boundary phase is non-magnetic to minimize grain-to-grain magnetic coupling. This keeps H_c high and also reduces noise in playback [6.37].

Rare-earth-containing amorphous alloys are of importance in perpendicular magneto-optic recording media such as GdCo, TbFe and mixtures of these two. In such applications the absence of grain boundaries, high electrical resistivity, strong Kerr rotation, and good mechanical properties combine with the ability to tailor their magnetic properties (M_s, K_u, T_c, T_{comp}) with continuous composition variations over broad ranges to provide distinct advantages over crystalline alloys.

6.6.4 Permanent Magnets

Amorphous alloys are not used, per se, as permanent magnets but can be a precursor phase suitable for the development of a segregated nanostructure suitable for hard magnetic materials. FeNdB alloys having compositions close to the high anisotropy phase $Nd_2Fe_{14}B_1$ can be melt spun either directly to a hard magnetic nanocrystalline phase (underquenched) or fully quenched to form a metallic glass having low coercivity. This amorphous phase can then be heat treated to develop the fine grain structure with a 2-14-1 nanocrystalline phase separated by non-magnetic Nd-rich and B-rich phases [6.38]. The physical principal for magnetic hardening here is similar to that in Co–Cr-based thin-film recording media. High anisotropy, single-domain grains separated by a non-magnetic grain-boundary phase is desired. If excess iron is present in the starting amorphous alloy, upon heat treatment to produce a nanocrystalline $Fe_{14}Nd_2B$ phase, the intergranular phase becomes enriched in iron and may be magnetic. A magnetic grain boundary phase can enhance the remanent induction of the magnet [6.39]. If too much iron is present, the hard nanocrystalline grains become exchange coupled to each other and as a result, the coercivity decreases.

6.7 Conclusion, Outlook

Rapidly solidified alloys exhibit a range of properties that challenge our ability to process and understand materials. Further, they offer a variety of new technical opportunities in diverse magnetic applications. Their fundamental magnetic properties can be understood as consequences of their electronic structure which in turn reflects their short-range chemical and topological order. Short-range order is a consequence of the chemical interactions between the constituents of a material. The structure over a range greater than a few nanometers reflects the processing conditions. Many of the technical advantages of metallic glasses stem from the fact that these materials are quenched from the melt: most impurities remain in metastable solid solution, there are no grain boundaries or precipitates, little segregation, and a broad range of compositions can be fabricated by the same process yielding a continuous spectrum of property values. The electrical resistivity is high for metallic alloys, suppressing eddy currents in ac applications. Further, the exchange interaction tends to average out any local magnetic anisotropy insuring soft magnetic characteristics in many alloy compositions. Magnetostriction remains one of the major factors inhibiting easy magnetization in metallic glasses.

Nanocrystalline materials derive their unique structure and hence properties from the high volume fraction of grain boundary phase which modulates the way the magnetic grains interact with each other. The homogeneous amorphous state is an advantageous starting point for development of nanocrystalline alloys.

Opportunities for future work include improved understanding of how the short-range order is established in the presence of rapid solidification. To what extent does it represent only the SRO that exists in the melt or has it developed beyond that during rapid solidification? There are opportunities for improved understanding of the electronic origins of local magnetic anisotropy and magnetostriction in metallic glasses. These properties are still not well understood even in crystalline alloys.

It is hoped that these challenges to our fundamental understanding of amorphous and nanocrystalline magnetism will be addressed as expanding technical applications place heavier demands on our ability to design and process superior magnetic materials.

References

6.1 S. Mader, A. Nowick: Appl. Phys. Lett. **17**, 57 (1965)
6.2 C.C. Tsuei, P. Duwez: J. Appl. Phys. **37**, 435 (1966)
6.3 P. Duwez, S. Lin: J. Appl. Phys. **38**, 4096 (1967)
6.4 R.C. O'Handley: Fundamental magnetic properties, in *Amorphous Metallic Alloys*, ed. by F.E. Luborsky (Butterworths, London 1983) p. 257
6.5 R.C. O'Handley: J. Appl. Phys. **62**, R15 (1987)
6.6 C.L. Chien, K.M. Unruh: Phys. Rev. Lett. B **24**, 1556 (1981)
6.7 R. Hasegawa, R. Ray: J. Appl. Phys. **49**, 4174 (1978)
6.8 R.C. O'Handley: Phys. Rev. B **18**, 930 (1978)
6.9 R.C. O'Handley, L.I. Mendelsohn, E.A. Nesbitt: IEEE Trans. MAG-**12**, 942 (1976)
6.10 K. Inomata, T. Kobayashi, M. Hasegawa, T. Sawa: J. Magn. Magn. Mater. **31–34**, 1577 (1983)
6.11 N. Hasegawa, M. Saito, N. Kataoka, H. Fujimori: J. Mater. Eng. Perf. **2**, 181 (1993)
6.12 G. Herzer: J. Magn. Magn. Mater. **157**, 133 (1996)
6.13 Polycrystalline permalloy, L.E. Tanner; amorphous FeBSi, R.J. Celotta, J. Unguris, M.J. Scheinfein, D.T. Pierce: Personal commun. (1997)
6.14 H. Unguris, R.J. Celotta, D.T. Pierce: Phys. Rev. Lett. **67**, 140 (1991)
6.15 D.I. Paul: J. Appl. Phys. **53**, 1649 (1982)
6.16 R.C. O'Handley, J. Megusar, S.-W. Sun, Y. Hara, N.J. Grant: J. Appl. Phys. **57**, 3563 (1985)
6.17 H.S. Chen, S.D. Ferris, H.J. Leamy, R.C. Sherwood, E.M. Gyorgy: Appl. Phys. Lett. **26**, 405 (1975)
6.18 Y. Imry, S.-K. Ma: Phys. Rev. Lett. **35**, 1399 (1975)
6.19 R. Alben, J.I. Budnick, G.S. Cargill III: In *Metallic Glasses*, ed. by J.J. Gilman, N.J. Leaney (Am. Soc. Metals, Metals Park, OH 1978)
6.20 F. Reif, *Fundamentals of Statistical and Thermal Physics* (McGraw-Hill, NY 1965) pp. 4–40
6.21 P. Oelhafen, E. Hauser, H.-J. Güntherodt, K. Benneman: Phys. Rev. Lett. **43**, 1134 (1979)
6.22 V.L. Moruzzi, P. Oelhafen, A.R. Williams, R. Lapka, H.-J. Güntherodt: Phys. Rev. B **27**, 2049 (1983)
6.23 P. Oelhafen, E. Hauser, H.-J. Güntherodt: Solid State Commun. **35**, 1017 (1980)
6.24 A. Amamou, G. Krill: Solid State Commun. **23**, 1087 (1980)
6.25 R.P. Messmer: Phys. Rev. B **23**, 1616 (1981)
6.26 T. Fujiwara: J. Non-Cryst. Solids **61&62**, 1039 (1984)
6.27 A. Collins, R.C. O'Handley, K.H. Johnson: Phys. Rev. B **38**, 3665 (1988)

6.28 K.H.J. Buschow: J Physique C **13**, 563 (1989)
6.29 H.J. Güntherodt, H.V. Künzi: In *Metallic Glasses*, ed. by J.J. Gilman, H.J. Leany (ASM, Metals Park, OH 1978) p. 247
6.30 R.C. O'Handley: Phys. Rev. B **18**, 2577 (1978)
6.31 G. Fish: Proc. IEEE **78**, 947 (1990)
6.32 F.E. Luborsky (ed.): *Amorphous Metallic Alloys* (Butterworths, London 1983)
6.33 R.C. O'Handley, C.-P. Chou, N. Cristofaro: J. Appl. Phys. **50**, 3603 (1979)
6.34 A. Datta, N.J. De Cristofaro, L.A. Davis: In *Proc. of Rapidly Quenched Metals IV*, ed. by T. Masumoto, K. Suzuki, Jpn. Inst. Metals (Sendai) **2**, 107 (1982)
6.35 R.C. O'Handley: J. Met. Eng. Perf. **2**, 211 (1993)
6.36 R. Schafer, W. Ho, J. Yamasaki, A. Hubert, F.B. Humphrey: IEEE Trans. MAG-**27**, 3678 (1991)
6.37 J.H. Judy: Mater. Res. Bulletin **15**, 63 (1990)
6.38 J.J. Croat, J.F. Herbert, R.W. Lee, F.E. Pinkerton: J. Appl. Phys. **55**, 2078 (1984)
M. Sagawa, M. Fujimura, M. Togawa, Y. Matsuura: J. Appl. Phys. **55**, 2083 (1984)
M. Sagawa, S. Hirozawa, H. Yamamoto, S. Fujimura, Y. Matsuura: Jpn. J. Appl. Phys. **26**, 785 (1987)
6.39 E. Kneller, R. Hawig: IEEE Trans. MAG-**27**, 3588 (1991)

General Reading

Beck H., H.-J. Güntherodt (eds.): *Glassy Metals II*, Topics Appl. Phys., Vol. 53 (Springer, Berlin, Heidelberg 1983)
Beck H., H.-J. Güntherodt: *Glassy Metals III*, Topics Appl. Phys., Vol. 72 (Springer, Berlin, Heidelberg 1994)
Gilman J.J., H.J. Leamy (eds.): *Metallic Glasses* (ASM, Metal Park, OH 1978)
Güntherodt H.J., H. Beck (eds.): *Glassy Metals I*, Topics Appl. Phys., Vol. 46 (Springer, Berlin, Heidelberg 1981)
Hasegawa R. (ed.): *Metallic Glasses: Magnetic, Chemical and Structural Properties* (CRC Press, Boca Raton, FL 1983)
Hasegawa R., R.A. Levy (eds.): *Amorphous Magnetism II* (Plenum, New York 1977)
Hooper H.O., A.M. de Graaf (eds.): *Amorphous Magetism I* (Plenum, New York 1973)
Liebermann H.H.: *Rapidly Solidified Alloys* (Dekker, New York 1993)
Moorjami K., J.M.D. Coey (eds): *Magnetic Glasses* (Elsevier, Amsterdam 1984)

7 Chemical Properties of Amorphous Alloys

K. Hashimoto

Let us consider a binary titanium–aluminum alloy. When aluminum is added to titanium up to 11 at.%, a single phase having the same structure as titanium, that is, the α-Ti phase is formed. A further increase in the aluminum content of the alloy results in the formation of two phase mixtures, such as α-Ti–Ti$_3$Al, Ti$_3$Al–TiAl, TiAl–TiAl$_3$ and TiAl$_3$–Al. Among them Ti$_3$Al, TiAl and TiAl$_3$ phases are called intermetallic compounds. As can be seen from this example, it is difficult to form a single-phase solid solution in a wide composition range when the alloy is prepared by conventional methods to form the equilibrium phases.

By contrast, for instance, when amorphous binary Ti–Al alloys are formed, they consist of a single-phase solid solution in a wide composition range. Although the amorphizable composition range depends on the alloy-preparation method, if the sputter deposition method, which will be explained later, is applied, Ti–Al alloys with about 40–70 at.% aluminum become a single phase of amorphous solid solution.

If one wishes to obtain alloys with specific properties by alloying, one generally has to succeed in preparing a single-phase solid solution in which all alloy constituents distribute homogeneously. As mentioned above, even for alloys with only two components of titanium and aluminum it is impossible to form a single solid solution phase except for the α-Ti phase, when a conventional method is used for alloy preparation.

The most attractive characteristic of the amorphous alloy is the single-phase nature forming a solid solution exceeding the solubility limit at equilibrium. This characteristic enables us to tailor new alloys with unique and useful properties.

7.1 Corrosion-Resistant Alloys in Aqueous Solutions

7.1.1 High Corrosion Resistance of Amorphous Fe–Cr Alloys

Except for precious metals, such as gold and platinum, ordinary metals found in nature are in the form of their oxidized states, such as oxides, sulfides, salts, etc. This fact indicates that for ordinary metals, the oxidized state is naturally more stable than the metallic state. In other words, in the atmosphere with some humidities at ambient temperature and pressure, metals which were separated artificially from oxygen wish to go back to the mine where they were stably present for a long time in the oxidized state. When metals are exposed to environments such as aqueous acids, they are dissolved as ions, and then bound to oxygen, forming oxides and/or hydroxides. This is the principle of corrosion of metals in aqueous environments.

In general, metals which are artificially reduced to the metallic state are readily oxidized in the atmosphere. If the oxidation product, such as oxide, covers the metal surface and separates the underlying metal surface from the environment, no further oxidation proceeds. This is the way to utilize metals which are readily oxidized. The thickness of the oxidation product is often very thin, such as several nanometers. The oxidation-product film is called passive film; and if the passive film is spontaneously formed by immersion of a metal into a corrosive solution, this situation is called spontaneous passivation. Consequently, if a metal is capable of forming the passive film spontaneously in an environment, the metal has high corrosion resistance and can be used in that environment.

If iron is immersed, for example, in 1 M HCl solution, iron is dissolved at first as ferrous (Fe^{2+}) ions and then forms iron rust, which mostly consists of the ferric (Fe^{3+}) ion, O^{2-} and OH^-. In this manner, almost all iron-based alloys including stainless steels, which are especially designed to avoid corrosion, suffer serious corrosion in the 1 M HCl solution.

In the late 1960s, it was found that amorphous iron–metalloid alloys possess surprisingly high mechanical strength with sufficient toughness. However, they are readily rusted in the atmosphere. In order to improve the corrosion resistance of amorphous iron–metalloid alloys with high strength, we tried to add chromium which is the most important alloying element to convert iron into stainless steels with higher corrosion resistance than iron.

This attempt gave us a surprising result [7.1]. Figure 7.1 is an example of the corrosion test. Since chromium is more active than iron, the addition of chromium to crystalline iron results in an increase in the corrosion rate of alloys without protective film formation. By contrast, the addition of chromium to amorphous iron–metalloid alloys leads to a decrease in the corrosion rate, and the amorphous alloys with 8 at.% or more chromium are immune to corrosion. Since then, a variety of extremely corrosion-resistant alloys have been found, and some reviews of corrosion resistance of amorphous alloys have been published [7.2–5].

7.1.2 Factors Determining the High Corrosion Resistance of Amorphous Alloys

a) High Activity of Amorphous Alloys

The high corrosion resistance found for amorphous alloys is based on spontaneous passivation even in very aggressive environments. They are spontaneously

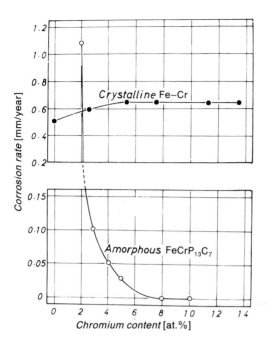

Fig. 7.1. Corrosion rates of amorphous Fe–Cr–13 P–7 C alloys and crystalline Fe–Cr alloys in 0.5 M NaCl as a function of the chromium content of the alloy

covered with a protective passive film whose cationic composition is not the same as the alloy composition, but beneficial elements enhancing the corrosion resistance are significantly concentrated. For instance, immersion of amorphous Fe–10 Cr–13 P–7 C alloy in 1 M HCl for 1 week gives rise to the formation of the passive film consisting exclusively of hydrated chromium oxyhydroxide, $CrO_x(OH)_{3-2x}$ [7.6]. In this manner, the chromium-enriched passive film is generally formed on alloys whose corrosion resistance is based on the presence of chromium whether the alloys are amorphous or crystalline. However, amorphous alloys are superior to crystalline counterparts in this regard. Table 7.1

Table 7.1. The cationic fraction of Cr^{3+} in passive films formed on amorphous chromium-containing alloys and stainless steels in 1 M HCl

Alloys	Cationic fraction of Cr^{3+} in passive films	Passivation process
Amorphous alloys		
Fe–10 Cr–13 P–7 C	0.97	Spontaneous passivation
Fe–3 Cr–2 Mo–13 P–7 C	0.57	Anodic polarization
Co–10 Cr–20 P	0.95	Spontaneous passivation
Ni–10 Cr–20 P	0.87	Spontaneous passivation
Ferritic stainless steels		
Fe–30 Cr(–2 Mo)	0.75	Anodic polarization
Fe–19 Cr(–2 Mo)	0.58	Anodic polarization

shows the concentration of Cr^{3+} ions in the passive film formed on chromium-containing amorphous and crystalline alloys in 1 M HCl [7.2]. Crystalline ferritic stainless steels which are not spontaneously passive in 1 M HCl are passivated by anodic polarization. As can be seen, the passivation of the ferritic stainless steels is based on the formation of the chromium-enriched passive film. Chromium is concentrated to 75 and 58 at.% in the passive films on Fe–30 Cr and Fe–19 Cr steels, respectively. The chromium enrichment is more significant in the passive films on amorphous alloys. The passive film on the Fe–10 Cr–13 P–7 C alloy consists exclusively of Cr^{3+} ions. Only 3 at.% of cations are Fe^{2+} and Fe^{3+} ions and 97 at.% of cations are Cr^{3+} ions. The chromium content in the film formed on the low chromium Fe–3 Cr–2 Mo–13 P–7 C alloy is comparable to that in the film on the ferritic Fe–19 Cr steel.

When we analyzed the underlying alloy surface, we found that the composition of the underlying alloy surface is not largely different from the bulk alloy composition. This indicates that the chromium-enriched passive film is formed as a result of preferential dissolution of other alloy constituents not necessary for passive film formation, while chromium oxyhydroxide is a very stable solid in 1 M HCl. This fact leads us to the conclusion that, the higher the reactivity of the alloy, the faster and higher the enrichment of the beneficial element in the passive film. Amorphous alloys are metastable and react rapidly with aqueous solutions. This leads to rapid dissolution of unnecessary alloy constituents and to fast and significant enrichment of beneficial Cr^{3+} ions in the surface film. Consequently, the film formation occurs without serious dissolution which is often observed for crystalline alloys even if passivation occurs.

In connection with a high activity of amorphous alloys there is an interesting experimental result. *Huerta* and *Heusler* [7.7] examined the activity of Co–25 at.% B alloy in amorphous and crystalline states. The Co–25 at.% B alloy does not contain elements forming a stable passive film in acids, and hence, the dissolution rate of the alloy in an acid reflects the reactivity of the alloy. They conducted heat treatment for 5 h at different temperatures. The results are shown in Fig. 7.2. The amorphous Co–25 at.% B alloy begins to crystallize at temperatures higher than 523 K forming a Co_3B phase in the amorphous matrix. The Co_3B phase has the same composition as the amorphous matrix. The formation of the crystalline phase with the same composition as the amorphous phase results in a decrease in anodic dissolution current. This reveals that the amorphous phase is more active than the crystalline counterpart. At 623 K where the amorphous phase almost disappears, a single Co_3B phase alloy shows the lowest activity for alloy dissolution. At further higher temperatures, the Co_3B phase is decomposed to Co and Co_2B phases by the disproportionation reaction. The formation of heterogeneous three-phase structure leads to an increase in the alloy dissolution rate.

As can be seen from the above example, the amorphous alloy is more active than the crystalline counterpart. If the alloy containing a strongly passivating element is more reactive in an aqueous solution, unnecessary elements for passivation dissolve rapidly, leaving the strongly passivating element on the alloy

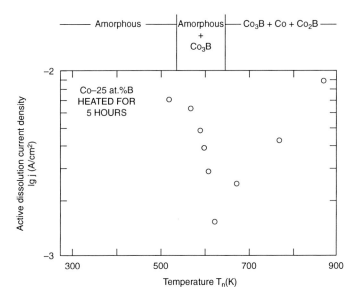

Fig. 7.2. Active dissolution current density of amorphous Co–25 at.% B alloy at 0.3 V (SCE) in 0.5 M SO_4^{2-} solution at pH 1.8 after heat treatments at various temperatures for 5 h

surface, and hence the passive film rich in the strongly passivating element is readily formed without serious corrosion of the alloy. This is partly responsible for high corrosion resistance of amorphous alloys containing a passivating element.

b) Homogeneous Nature of Amorphous Alloys

Because of their amorphous nature, amorphous alloys are free of crystalline defects such as grain boundaries, dislocations, stacking faults, etc. If a single amorphous phase alloy is formed, the alloy consists of a single-phase solid solution exceeding the solubility limit at equilibrium, and is free of chemical heterogeneity. Accordingly, the amorphous alloys are regarded as the ideally homogeneous alloys.

The chemical heterogeneity of an alloy is often responsible for the formation of anodic and cathodic sites on the alloy and for the high corrosion rate. For instance [7.8], crystalline iron and Fe–Mo alloys cannot be passivated by anodic polarization and dissolve actively in 1 M HCl, whereas amorphous Fe–Mo–13 P–7 C alloys are passivated by anodic polarization in 1 M HCl. We analyzed the passive film of a few nanometer thickness formed on the amorphous Fe–Mo–13 P–7 C alloys by X-ray photoelectron spectroscopy. X-ray photoelectron spectroscopy, because of a very high surface sensitivity, is useful to analyze a few nanometer thick passive film and the underlying alloy surface, and gives information on the valence of the film constituents along with the compositions of the film and the underlying alloy surface. The passive film formed on the amorphous Fe–Mo–13 P–7 C alloys in 1 M HCl by anodic polarization contains a very low concentration of Mo ions and consists exclusively of iron oxyhydroxide. On the other hand, crystalline iron cannot be passivated by anodic polarization in 1 M HCl but is passivated in 0.5 M H_2SO_4, forming a passive iron oxyhydroxide film which is the same as the passive film formed on the amorphous Fe–Mo–13 P–7 C alloys in 1 M HCl. In spite of the fact that the passive iron oxyhydroxide itself is not unstable in 1 M HCl, iron cannot be passivated in 1 M HCl. Figures 7.3 and 4 illustrate how effective the homogeneous nature of the alloy is in providing the high corrosion resistance based on passivation.

c) Beneficial Effect of Phosphorus in Amorphous Alloys

Phosphorus is well known to segregate at grain boundaries of stainless steels and to induce intergranular corrosion and stress-corrosion cracking of the stainless steels. However, phosphorus contained uniformly in amorphous alloys is effective in decreasing the corrosion rate and in enhancing passivation of amorphous alloys with a passivating element.

The corrosion rate of amorphous Ni–P alloys, for instance, in 1 M HCl is significantly lower than crystalline nickel. Kinetic and analytical investigations [7.9] revealed the role of phosphorus as follows. Immersion of the amorphous Ni–P alloys in 1 M HCl gives rise to preferential dissolution of nickel, leaving

Fig. 7.3. Amorphous Fe–Mo–13 P–7 C alloys are chemically homogeneous and can be covered uniformly by a passive iron oxyhydroxide film even in 1 M HCl

Fig. 7.4. Although a passive iron oxyhydroxide film is not unstable in acidic chloride solutions such as 1 M HCl, the stable passive film cannot be formed on chemically heterogeneous sites of crystalline iron and hence catastrophic corrosion occurs

elemental phosphorus on the alloy surface. The elemental phosphorus is accumulated and forms the surface layer which can act as the barrier layer against diffusion of nickel through the layer to dissolve into the solution. Accordingly, dissolution of nickel leads to thickening of the elemental phosphorus barrier layer and to the decrease of the dissolution current. In this situation, the decrease of the dissolution current with time at a constant potential follows Fick's second law. A typical example is shown in Fig. 7.5. The equation written in the figure is based on Fick's second law, and the reciprocal of the current density squared changes linearly with polarization time.

Phosphorus has another beneficial effect in enhancing the corrosion resistance. Let us consider reactions by which metals are corroded. As can be seen from the example that corrosion of zinc in a dilute sulfuric acid is accompanied by hydrogen evolution, corrosion of a metal is often accompanied by hydrogen evolution. These reactions can be expressed as:

$$Zn = Zn^{2+} + 2e, \tag{7.1}$$

$$2H^+ + 2e = H_2. \tag{7.2}$$

Equations (7.1) and (7.2) are called anodic and cathodic reactions, respectively. Another cathodic reaction in aqueous solutions is the reduction of oxygen dissolved into the solutions:

$$O_2 + 2H_2O + 4e = 4OH^-. \tag{7.3}$$

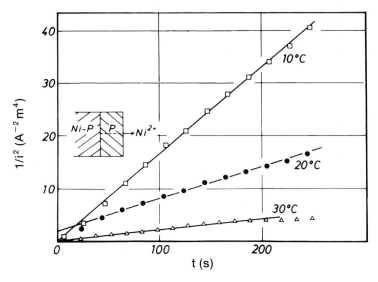

Fig. 7.5. Reciprocal of the square of the current density i for dissolution of nickel from a melt-spun amorphous Ni–19 at.% P alloy specimen polarized at 0.1 V (SCE) in 1 M HCl at three different temperatures as a function of time t ($1/i^2 = V^2\pi(t+t_0)/[(nFa_i)^2 D]$; V: molar volume, a_i: bulk mole fraction of the dissolving component)

These cathodic reactions are the counter reaction to metal dissolution. In the solution, an increase in the positive charge by dissolution of metal ions is canceled by consumption of protons or by formation of hydroxyl ions by the cathodic reaction. Similarly, electrons left in the metal as a result of dissolution of metal ions are consumed by the reduction of protons or dissolved oxygen. In this manner, these reactions proceed to maintain the electroneutrality in both the solution and metal. In other words, the dissolution of metals can occur only when the electron consumption in the cathodic reaction proceeds at the same rate as the rate of electron release in the metal dissolution.

Chromium dissolves in the form of Cr^{2+} ions and, if the oxidizing power is further high, chromium is oxidized to Cr^{3+} ions. Cr^{3+} ions constitute a stable solid film with oxygen and hydroxyl ions even in strong acids in the form of $CrO_x(OH)_{3-2x}$. This is the passive film of chromium. The passive film formation is also the anodic reaction which requires the faster oxygen reduction than that in the metal dissolution. Under natural conditions where no electricity is supplied by an outer circuit, the higher oxidizing power, that is, the faster oxygen reduction, corresponds to the fact that the cathodic reactions, particularly the oxygen reduction, occur readily. Since the concentration of oxygen dissolved in the solution is proportional to the atmospheric pressure in equilibrium with the solution, the rate of reaction (7.3) is mostly dependent on the activity of the surface for this reaction.

The elemental phosphorus layer is particularly active for the oxygen reduction (7.3) [7.10]. Accordingly, for amorphous alloys containing phosphorus and a strongly passivating element such as chromium, the formation of the elemental phosphorus layer not only prevents alloy dissolution but also accelerates the passive film formation, that is, spontaneous passivation owing to the high activity for reaction (7.3).

7.1.3 Recent Efforts in Tailoring Corrosion-Resistant Alloys

One serious restriction for practical utilization of corrosion-resistant amorphous alloys prepared by melt-spinning methods is their limited thickness of several tens of microns, because the thickness of the melt which can be rapidly quenched from the liquid state for the formation of the amorphous structure is limited. Furthermore, conventional welding techniques cannot be applied, because heating of amorphous alloys leads to crystallization and to a loss of their superior characteristics based on the amorphous structure.

One of the solutions to this problem is the preparation of amorphous surface alloys having the specific characteristics on conventional crystalline bulk metals. Tailoring new corrosion-resistant surface alloys has recently been performed mostly by the sputter deposition technique. This technique is known to form a single amorphous phase alloy in the widest composition region among the various methods for amorphization.

a) Aluminum-Refractory Metal Alloys

Aluminum is the most widely used metal next to iron. Aluminum is not highly corrosion resistant and corrodes in both acidic and alkali environments. Alloying with refractory metals is a potential method in enhancing the corrosion resistance [7.11]. However, the boiling point of aluminum is sometimes lower than the melting points of refractory metals. Accordingly, the conventional casting methods cannot be applied to the preparation of aluminum-refractory metal alloys. This bottle neck has been overcome by utilizing sputter deposition techniques for alloy preparation. The sputter deposition technique does not rely on melting to mix the alloying constituents and hence is suitable of forming a single-phase solid solution even when the boiling point of one component is lower than the melting point of the other components and/or when one component is immiscible with another component in the liquid state.

The sputter deposition method is applied in tailoring corrosion-resistant aluminum alloys. Figure 7.6 shows the structure of various binary aluminum alloys. These alloys have been successfully prepared in a single amorphous phase in wide composition ranges. Alloying is very useful in enhancing the corrosion resistance. Figure 7.7 shows a comparison of corrosion rates of the various aluminum alloys with those of conventional corrosion-resistant alloys measured in 1 M HCl at 30 °C. The use of aluminum alloys in 1 M HCl has not been previously considered. However, when aluminum is alloyed with various

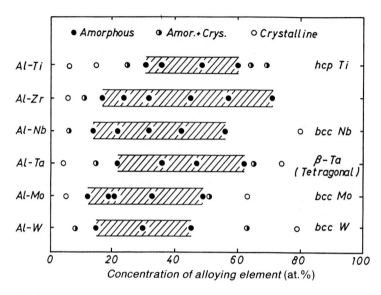

Fig. 7.6. Structure of sputter-deposited aluminum alloys identified by X-ray diffraction

refractory elements, the alloys possess sufficiently high corrosion resistance even in 1 M HCl. Except for Al–Ti alloys, the corrosion resistance in 1 M HCl increases with increasing alloying additions. Al–Ti alloys dissolve actively, but other amorphous aluminum alloys are spontaneously passive even in 1 M HCl. Amorphous Al–Ta and Al–Nb alloys are especially corrosion resistant.

b) Chromium-Refractory Metal Alloys

Titanium, zirconium, niobium, tantalum as well as aluminum are called valve metals, because, when immersed in a suitable electrolyte, they readily pass current when the valve-metal electrode is polarized cathodically, whereas when this electrode is made the anode, the current passed is very small and often time dependent. The valve metals except for aluminum are all passivated in strong acids. Chromium also has a very strong passivating ability. Consequently, if one succeeds in preparing chromium alloys with these valve metals, they seem to be ideal corrosion-resistant alloys in aqueous environments [7.12]. We tried to produce these alloys by sputtering. Figure 7.8 shows the structure of sputter-deposited chromium-valve metal alloys as a function of alloying additions. These alloys show an amorphous single solid solution structure over wide composition ranges. These were all new amorphous alloys. Their corrosion resistance to concentrated hydrochloric acids is remarkably high.

Figure 7.9 shows the change in the corrosion rate of Cr–Ti and Cr–Zr alloys in 6 M HCl solution at 30 °C and Cr–Nb and Cr–Ta alloys in 12 M HCl solution at 30 °C as a function of the valve-metal content of the alloy. In

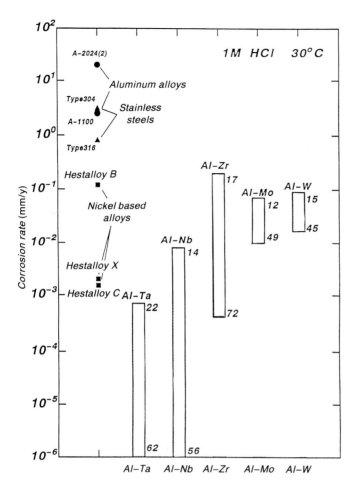

Fig. 7.7. Corrosion rates of various aluminum alloys and conventional corrosion-resistant alloys measured in 1 M HCl at 30 °C

6 M HCl solution chromium and titanium dissolve actively. However, the Cr–Ti alloys show very low corrosion rates which are several orders of magnitude lower than those of alloy components. Binary Cr–Zr alloys also show low corrosion rates. In spite of the fact that the corrosion rate of chromium is five orders of magnitude higher than that of Zr, the corrosion rate of the alloys decreases with increasing chromium content of the alloy. Amorphous Cr–Nb and Cr–Ta alloys show very high corrosion resistance which is higher than that of the alloy components. These results indicate that, if both components of binary alloys have a strong passivating ability, the alloys are able to possess better corrosion resistance than the alloy components. The corrosion rate of Cr–Ti, Cr–Zr and Cr–Nb alloys tends to decrease with chromium content. The

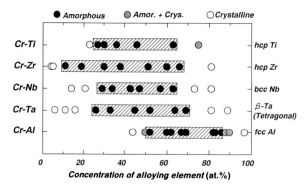

Fig. 7.8. Structure of sputter-deposited chromium-valve metal alloys as a function of alloying additions

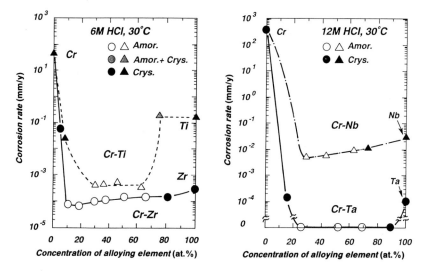

Fig. 7.9. Change in the corrosion rate of Cr–Ti and Cr–Zr alloys in 6 M HCl solution at 30 °C and Cr–Nb and Cr–Ta alloys in 12 M HCl solution at 30 °C

corrosion rates of Cr–Ta alloys are extremely low and are lower than the level measurable by Inductively Coupled Plasma (ICP) spectrometry, 2×10^{-5} mm per year. It can therefore be said that the amorphous Cr–Ta alloys have the highest corrosion resistance among metallic materials so far produced in the world. These alloys are all passivated spontaneously, forming the passive film consisting of both chromium and valve-metal cations.

An interesting fact has been found with regard to the binding energy of core electrons of elements constituting the passive film. Figure 7.10 shows the correlation of the binding energies of the Cr^{3+} $2p_{3/2}$ and Ta^{5+} $4f_{7/2}$ electrons with the cationic fraction of tantalum in the surface film formed on the Cr–Ta alloys.

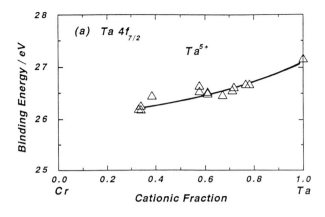

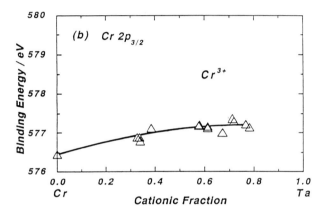

Fig. 7.10. The correlation of the binding energies of the Cr^{3+} $2p_{3/2}$ (a) and Ta^{5+} $4f_{7/2}$ (b) electrons with the cationic fraction of tantalum in the surface film formed on Cr–Ta alloys

The formation of the film containing both tantalum and chromium results in charge transfer from chromium to tantalum in the film. Similar charge transfer from chromic ions, Cr^{3+}, to valve-metal cations has been found for Cr–Ti, Cr–Zr and Cr–Nb alloys.

The charge transfer between different cations indicates that these cations are located very closely to show the electronic interaction. This means that the passive film does not consist of a simple mixture of chromium oxyhydroxide and valve-metal oxide, but is composed of a double oxyhydroxide of chromium and valve-metal cations. The resultant double oxyhydroxide films are more protective than valve-metal oxide films in these aggressive solutions, and increasing the chromium content of the alloys increases the chromium content of the double-oxyhydroxide film and increases the corrosion resistance of the binary chromium-valve metal alloys.

As can be seen from these examples, sputter deposition is the potential technique in tailoring new corrosion-resistant alloys, and hence the study is in progress [7.13].

7.2 Corrosion-Resistant Alloys at High Temperatures

Another interesting fact is the extremely high corrosion-resistance of Al-refractory metal alloys at high-temperature corrosion [7.14] in sulfidizing and oxidizing environments. The high temperatures of metallic materials in sulfur-containing atmospheres are much more severe than in purely oxidizing environments. All conventional oxidation-resistant alloys suffer catastrophic corrosion in sulfur-containing atmospheres at high temperatures, because of the poor protective properties of sulfide scales. For instance, the non-stoichiometry of sulfide scales formed on these alloys often reaches 10%. Because of rapid diffusion of cations through the defective sulfide scale they are sulfidized very rapidly.

Some refractory metals such as molybdenum and niobium are, however, resistant to sulfide corrosion and their sulphidation rates are almost comparable to the oxidation rate of chromium. These metals, however, have very low resistance against high-temperature oxidation in spite of the fact that sulfidizing atmospheres in industry are often oxidizing. On the other hand, the best alloying element to form a protective scale in oxidizing environments is aluminum, and the second best is chromium. These metals form alumina and chromia scales, respectively. We, therefore, thought that aluminum-refractory metal alloys must be the best materials having high resistance against both oxidation and sulfidation.

The corrosion rate at high temperatures is evaluated by the rate constant. If the protective scale is formed by high-temperature corrosion, the slowest step of corrosion, that is, the rate-determining step is diffusion of cations or anions through the reaction-product scale. Under this condition, a parabolic increase in the corrosion-weight gain is observed. Namely, the square of the weight gain, Δw, becomes proportional to the reaction time, t.

$$(\Delta w)^2 = kt, \tag{7.4}$$

where k is the parabolic rate constant. The corrosion rate at high temperatures is generally compared to the parabolic rate constant.

Figure 7.11 shows sulfidation-(solid lines) and oxidation-(dotted lines)-rate constants for amorphous Al–Mo and Al–Mo–Si alloys as well as several high-temperature alloys. As can be seen from the comparison between solid and dotted lines, the sulfidation rate of conventional oxidation-resistant crystalline alloys is generally many orders of magnitude higher than the oxidation rate. By contrast, the sulfidation rates of Al–Mo and Al–Mo–Si alloys are significantly low and comparable to the oxidation rate of oxidation-resistant alloys. Furthermore, the sulfidation-rate constants of these alloys are more than one order of

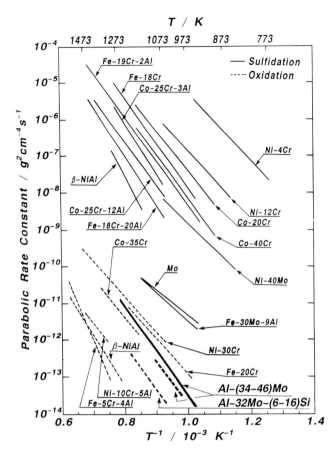

Fig. 7.11. Temperature dependence of the sulfidation (*solid lines*) and oxidation (*dotted lines*) rates for Al–Mo and Al–Mo–Si alloys and analogous dependence of the rates for several binary and ternary alloys

magnitude lower than those of pure molybdenum, and molybdenum-containing Fe–30 Mo–9 Al alloys. This result is of particular importance because this is the first time in the history of corrosion science that a metallic material has been obtained whose corrosion rate in highly sulfidizing atmosphere is comparable to the oxidation rate of oxidation-resistant alloys.

The high sulfidation resistance of the Al–Mo alloys is attributed to the formation of the MoS_2 phase, constituting the major part of the inner-barrier layer of the scale, although the Al_2S_3 outer layer is formed. The better protective properties of the sulfide scale formed on the Al–Mo alloys in comparison with those of the MoS_2 scale on pure molybdenum result from a lower defect concentration in the aluminum-doped MoS_2 phase.

The oxidation rate of Al–Mo alloys is comparable to that of chromia-forming alloys although it is higher than that of alumina-forming alloys.

However, the oxidation rate at temperatures higher than 900 °C is very high. The scale consists mostly of alumina, but, because of high molybdenum contents of the alloys, molybdenum is also oxidized, forming volatile MoO_3. Since the melting point of MoO_3 is 793 °C, the formation of low-melting-point MoO_3 is responsible for the relatively low oxidation resistance of Al–Mo alloys. Accordingly, an attempt to improve the oxidation resistance was made by adding silicon to the Al–Mo alloys. The ternary Al–Mo–Si alloys have high sulfidation resistance similar to that of the Al–Mo alloys and have a higher oxidation resistance than Al–Mo alloys. This is attributable to the formation of molybdenum silicide, which is stable against oxidation. During sulfidation and oxidation, amorphous alloys are crystallized forming intermetallics. Al–Mo alloys form Al_8Mo_3 and Mo_3Al phases. The molybdenum-rich Mo_3Al phase is readily oxidized, forming volatile MoO_3. Accordingly, when the alumina scale surface on the Al–Mo alloys was analyzed, a low concentration of molybdenum was always found. By contrast, Al–Mo–Si alloys are crystallized to Al_8Mo_3 and Mo_5Si_3 phases without forming the easily oxidizable molybdenum-rich Mo_3Al phase. The Mo_5Si_3 phase is very stable against oxidation. Accordingly, any molybdenum and silicon were detected in the top-most surface of the alumina scale.

As shown in Fig. 7.6, there are a variety of aluminum-refractory metal alloys. Consequently, we can expect further to find alloys with high resistance against oxidation and sulfidation [7.15].

7.3 Electrodes for Electrolysis of Aqueous Solutions

7.3.1 Electrode Materials

The electrode is used for electrolysis of a solution. The electrode for an electrolytic reaction is required to ensure not only the electrocatalytic activity for the reaction but also the durability against the electrolytic conditions. For example, the anode for the electrolysis of hot concentrated sodium-chloride solutions is used for the production of chlorine in chlor-alkali industry:

On the anode

$$2\,Cl^- = Cl_2 + 2\,e; \tag{7.5}$$

On the cathode

$$2\,Na^+ + 2\,H_2O + 2\,e = 2\,NaOH + H_2. \tag{7.6}$$

For the production of chlorine in chlor-alkali industry, the anode and cathode compartments are separated in order to obtain chlorine and sodium hydroxide separately. On the other hand, seawater is often used in cooling systems in industrial plants and ships. In order to avoid clogging of seawater cooling systems, the sterilization of seawater is often carried out by adding sodium hypochlorite (NaClO) to the cooling seawater. The sodium hypochlorite can be formed by electrolysis of seawater. The anode and cathode reactions in the electrolysis of seawater are the same as those in chlor-alkali industry. However, for the production of sodium hypochlorite, the anode and cathode are not separated but are very closely spaced for the further reaction of chlorine with sodium hydroxide in seawater as:

$$2\,NaOH + Cl_2 = NaClO + NaCl + H_2O. \tag{7.7}$$

The anode materials for these reactions are required to have a high activity for the production of chlorine and a low activity for the evolution of parasitic oxygen, which unavoidably occurs in the electrolysis of aqueous solutions. Furthermore, the anode is required to have a very high corrosion resistance against highly oxidizing conditions producing nascent chlorine. The electrolytic conditions of seawater are particularly more severe than those in chlor-alkali industry, since the production of chlorine from natural seawater is more difficult than from hot concentrated sodium-chloride solutions at low pH.

The characteristics of the amorphous alloys, which can contain various elements exceeding the solubility limit at equilibrium, render them quite suitable for obtaining novel electrode materials by alloying. Since palladium has a very high activity for the electrolytic chlorine evolution and a low activity for oxygen evolution in dilute sodium-chloride solutions, and because it is quite readily corroded in sodium-chloride solutions, alloying and amorphization have been tried in order to improve the corrosion resistance of this metal without decreasing its electrocatalytic activity.

The amorphous palladium alloys have exhibited a good performance in chlorine production in hot concentrated sodium-chloride solutions. However, the activity of these alloys for the electrolysis of seawater was not sufficiently high, but a method for the activation of the amorphous palladium-based alloys was found. The activation method consists of electrodeposition of zinc on the amorphous palladium alloys, heat treatment at temperatures lower than the crystallization temperature of the amorphous alloys for the diffusion of zinc into the amorphous alloys, and leaching of zinc from the alloys by immersion in concentrated KOH solution. By this treatment, the surface of the amorphous alloys are once further alloyed with zinc and then zinc is selectively dissolved in the KOH solution. This treatment results in a significant increase in the effective surface area by surface roughening. However, the activation treatment cannot be applied to crystalline metals since preferential grain-boundary diffusion of zinc occurs and leads to grain-boundary degradation after leaching of zinc without increasing the activity.

Figure 7.12 shows an example of the high-current efficiency of the surface-activated amorphous alloy for chlorine production in a sodium chloride solution [7.16]. The commercial Pt–Ir/Ti electrode is known to have the highest activity among commercial anodes for the electrolysis of seawater. Although the

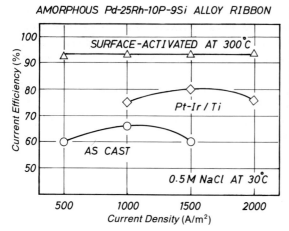

Fig. 7.12. Current efficiencies of as-prepared and surface-activated amorphous Pd–25 Rh–10 P–9 Si alloy and a commercial Pt–Ir alloy coated on Ti (Pt–Ir/Ti) electrode for chlorine evolution in 0.5 M NaCl solution at 30 °C

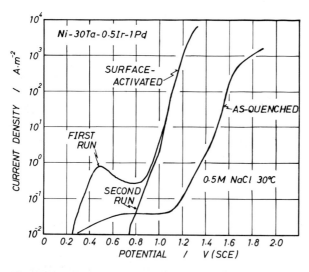

Fig. 7.13. Anodic polarization curves measured in 0.5 M NaCl solution at 30 °C for melt-spun amorphous Ni–30 Ta–0.5 Ir–1 Pd alloy specimens before and after the specimen surface was activated by immersion in HF

activity of the as-prepared amorphous alloy is not high, the surface-activated alloy shows a quite high efficiency for chlorine production.

However, since the palladium-based alloys are composed of precious metals, tailoring of inexpensive electrode was requested. The new electrode materials have been produced by a combination of electrocatalytically active elements and corrosion-resistant elements. An example is shown in Fig. 7.13. A sharp increase in the current density at potentials higher than 1 V (SCE) is due to rapid chlorine evolution [7.17]. Amorphous Ni–Ta and Ni–Nb alloys containing very small amounts of platinum-group elements show a very high corrosion resistance due to their passivation under anodic polarization conditions. Although these alloys were not active for chlorine evolution, an effective activation method was found. When these alloys are immersed in HF solutions, in which tantalum, niobium and nickel are preferentially dissolved, surface-enrichment of the platinum-group elements and surface roughening give rise to an almost four orders of magnitude increase in the activity for chlorine production. The current efficiency of the surface-activated Ni–Nb and Ni–Ta alloys containing a few at.% of iridium and/or palladium for chlorine evolution is higher than 90%.

It has been found that the same type of electrodes, with different compositions, are effective as the anodes for electrowinning of zinc.

7.3.2 Preparation of Electrodes

When melt-spun ribbon-shaped amorphous alloys are used as electrodes [7.18], the alloys have a high electric resistance based on the limited thickness. For instance, when an amorphous alloy of $100\,\mu\Omega\cdot$cm specific resistivity and $5\,\mu$m thickness is used as an anode with a high electrocatalytic activity, the anode input terminals must be connected every 2 cm of the length of the anode. Otherwise, the current density fluctuation cannot be maintained within 780–1000 A/m^2 due to voltage drop based on the electric resistance of the anode. It is, therefore, difficult to utilize high corrosion resistance and electrocatalytic activity of thin ribbon-shaped amorphous alloys.

One of the solutions to this problem is the preparation of amorphous surface alloys having the specific characteristics on conventional crystalline bulk metals. When an electrode consists of the amorphous surface alloy on a bulk crystalline substrate, the electrocatalytically active amorphous surface alloy is supported by the bulk metal which acts as the electric conductor during electrolysis. Consequently, the preparation of amorphous surface alloys was performed by using high-energy-density beam treatment [7.16].

Instantaneous irradiation of a metal surface by a high-energy-density beam such as a laser or electron beam is able to melt a small volume of the metal surface instantaneously. The heat of this melt can be rapidly absorbed by the large volume of cold solid metal surrounding the melt, with a consequent rapid quenching thereof. Accordingly, the application of high-energy-density-beam melting and subsequent self-quenching to a metal surface is one of the effective methods for rapid solidification. This technique can be applied to surface amorphization by choosing a proper surface composition. However, amorphization of a large surface area requires overlapping traverses by the high-energy-density beam, as shown in Fig. 7.14. The processing of a large surface area, therefore, requires heating of the previously amorphized phase for irradiation melting of a portion adjacent to the previously amorphized phase. As can be seen from the cooling sequence shown in Fig. 7.15, when the small volume 1 in the surface is instantaneously melted by the laser irradiation, the heat of the melt 1 is rapidly absorbed by a large volume of the directly surrounding solid metal. In this

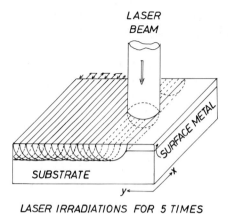

Fig. 7.14. Schematic of laser processing for the formation of a surface alloy on a bulk substrate

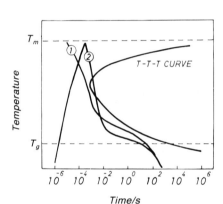

Fig. 7.15. A schematic drawing of the cooling sequence for laser surface treatment (T_m: melting point; T_g: glass transition temperature)

manner, once melted, the amorphization of a single laser trace occurs in relatively wide alloy compositions. However, when the laser beam is irradiated onto an area neighboring the previously amorphized phase, a portion of the previously amorphized region 2 is heated without melting by contact with the neighboring melt. The thermal history of the previously amorphized region 2 includes the heating curve during the laser irradiation to form the melt in the neighborhood. Hence, the cooling curve of the region 2 intersects the crystalline nose of the Temperature–Time-Transition (TTT) curve with the consequent formation of the crystalline phase in the region 2. Thus, the crystalline phase always appears in the heat-affected zone at the border of the neighboring laser irradiation traces. Consequently, high-energy-density-beam processing is the most difficult method for the preparation of thermodynamically metastable amorphous alloys. Nevertheless, the prevention of crystallization in the previously amorphized zone during processing by a high-energy-density beam can be achieved by proper selection of an alloy composition having a higher glass-forming ability, or by shortening the irradiation time by using a higher energy density.

In using laser processing, we succeeded in preparing extremely corrosion-resistant amorphous surface alloys on a conventional bulk mild steel by choosing the proper surface compositions, although the completely amorphizable composition range is seriously restricted. Electrodes for the electrolysis of aqueous solutions, such as palladium-based alloys on titanium and Ni–Nb–platinum-group metal alloys on niobium, were also prepared by laser processing. Figures 7.16 show a comparison of top views of crystalline and amorphous surface alloys after irradiation melting by high-energy-density beams.

0.1mm

Fig. 7.16. Top views of the surfaces of electron-beam processed Pd–25 Rh–10 P–9 Si alloy on titanium (**a**) and laser-treated type 304 stainless steel (**b**)

Crystalline metals always show shell-like patterns after laser-irradiation melting. On the contrary, an individual trace of a high-energy-density beam in the amorphized surface is very smooth just as an inorganic glass.

The laser processing is, however, rather slow, mostly because of the high reflectivity of the infrared CO_2-laser beam at 10.6 μm from solid surfaces of metals. Furthermore, it is difficult to oscillate the high-power laser beam focusing on the specimen's surface, and hence a heavy stage must be reciprocated, as shown in Fig. 7.14.

By contrast, an electron beam, another high-energy-density beam, is easily absorbed by metals. In addition, it is quite easy to oscillate the electron beam with a high frequency during irradiation of the slowly moving metals. Figure 7.17 schematically shows the electron-beam processing. The laser-processing conditions by which the amorphous surface alloys were prepared were used for the electron-beam processing on the basis of the assumption that more than 90% of the laser beam was reflected from the specimen's surface, whereas almost 100% of the electron beam is absorbed. The diameter of the electron beam was kept at about 0.4 mm by using double-condenser lenses to avoid a change in the beam diameter with oscillation. During irradiation by the electron beam, which oscillated along the x-axis at 100–600 Hz, the table on which the specimen was mounted was moved along the y-axis. After processing was completed from one end to the other of the specimen along the y-axis, the table was shifted along the x-axis and the direction of the movement of the table along the y-axis was reversed. Various amorphous surface alloys were prepared on conventional crystalline metals by electron-beam processing to obtain electrodes for the electrolysis of aqueous solutions.

A comparison between laser- and electron-beam processing gave an interesting result. For instance, it was found in preparing an amorphous Ni–Nb alloy on niobium that if the electron beam is assumed to be absorbed by the metal specimen with 100% efficiency, only 1.7% of the CO_2-laser beam is

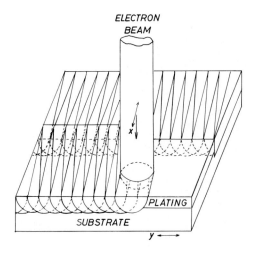

Fig. 7.17. Schematic of electron-beam processing for the formation of a surface alloy on a bulk substrate

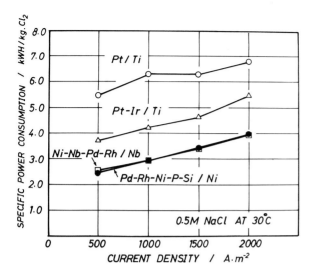

Fig. 7.18. Specific power consumption of amorphous surface alloys and commercial Pt–Ir/Ti and Pt/Ti anodes for chlorine evolution by electrolysis of a 0.5 M NaCl solution at 30 °C

absorbed. Accordingly, when 6 kW CO_2 laser- and 6 kW electron-beam machines were used for the formation of 1 m^2 of amorphous Ni–40 Nb alloy on niobium from a 0.015 mm thick nickel-plated niobium specimen, the processing times are about 8 h and 22 min, respectively. Consequently, for the preparation of a plane amorphous surface alloy, electron-beam processing is more convenient than laser processing.

The electron-beam-processed amorphous surface alloys are used as the electrodes for the production of sodium hypochlorite by the electrolysis of seawater. Figure 7.18 shows examples of the high efficiency of the amorphous alloy electrodes. This figure exhibits the electricity required for the production of 1 kg of Cl_2 as a function of current density. The Pt–Ir/Ti electrode is known to have the highest activity among currently used electrodes. Industrial electrolysis is generally carried out at 1000 A/m^2. The electricity required at 1000 A/m^2 for the amorphous alloy electrodes is almost two thirds of that for the Pt–Ir/Ti electrode. Accordingly, the electron-beam-processed amorphous surface alloy anodes are decidedly energy-saving anodes.

7.4 Catalysts for Prevention of the Greenhouse Effect and Saving the Ozone Layer

Amorphous alloys have a potential to produce new catalysts or precursors of new catalysts for special reactions, because they form single-phase solid solutions containing various effective elements exceeding the solubility limit in the equilibrium state. In general, the alloying with effective elements changes the electronic state of the alloy-constituent elements. It is also known that different reactant species sometimes coordinate preferentially to different elements on the catalyst surface. Accordingly, for amorphous alloys, new synergistic effects of different elements and modification of the electronic state of the catalyst elements can be expected.

Recently, particular emphasis has been devoted to use amorphous alloys as catalyst precursors rather than catalysts themselves, because oxidation and reduction strongly affect their physical and chemical properties. Controlled oxidation of amorphous alloys is known as an interesting technique for preparing supported metal catalysts with unique structure and morphology. The metal-supported oxides can be formed by selective oxidation of one or more alloy constituents.

We tried to tailor new catalysts from amorphous alloys for the prevention of a concentration increase of the greenhouse gases and saving the ozone layer in addition to abundant energy supply. Based on these investigations, we proposed the "green" materials research project [7.19, 20] consisting of four subjects as shown in Fig. 7.19. All these problems can be solved by tailoring novel materials having attractive and unique properties such as catalytic, electrocatalytic and corrosion-resistant properties.

7.4.1 CO_2 Recycling

There are a number of proposals to solve the CO_2 problem. However, since CO_2 is produced in energy-consuming processes, unless a new energy source is considered, the CO_2 problem cannot be solved. Our proposal of CO_2 recycling not only provides the solution of the atmospheric problems but also supplies

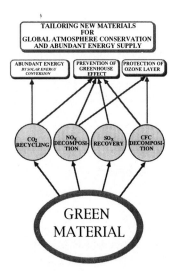

Fig. 7.19. Schematic diagram of the "green" materials research project

7.4 Catalysts for Prevention of the Greenhouse Effect and Saving the Ozone Layer

abundant energy, and it is based on the combination with solar-cell operation at deserts. When solar cells are operated with a 10% efficiency for 8 h/d by using the solar constant of 2 cal/cm^2 · min, in order to generate the electricity corresponding to the energy emitted by burning 6000 Mt/y of carbon in the whole world, we need 13.4 Mha of the area. This is only 0.83% of the area of the main deserts on the earth. It is, however, impossible to supply electricity over 1000 km and hence, we need another energy carrier. We can transmit the electricity generated by the solar-cell operation on the deserts to the nearest coast and produce H_2 by electrolysis of seawater. H_2 thus produced can be liquified and transported to fuel-burning plants by tankers. However, direct usage of H_2 has various disadvantages. Because of a lower energy content as the fuel, H_2 cannot be used directly for high-speed vehicles. Liquefaction consumes about one third of energy, and the energy content per volume is too low to be carried by liquefied H_2 tankers.

On the contrary, CH_4 has a 3.32 times higher energy per volume than H_2, and CO_2 is easily converted to CH_4 by the reaction with H_2. Figure 7.20 shows our proposal of the CO_2 recycling. The electric power is generated by the solar-cell operation at deserts and transmitted to the nearest coast. Using this electricity H_2 can be generated by the electrolysis of seawater. On the other hand, CO_2 is recovered at fuel-burning plants, liquefied and transported by tankers to the coast closest to the deserts. CH_4 can be produced by the reaction of CO_2 and H_2 and liquefied. The liquefied CH_4 can be transported by tankers to fuel-burning plants.

The problems which must be solved by material scientists in the CO_2 recycling are electrodes for electrolysis of seawater and catalysts for conversion of CO_2 to CH_4 by the reaction with H_2. In order to utilize the electricity generated in deserts where water is not available, not fresh-water but seawater

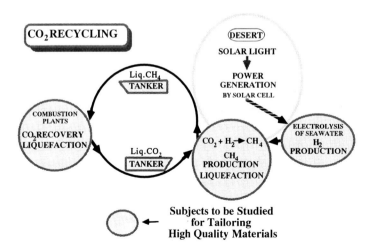

Fig. 7.20. Schematic diagram of CO_2 recycling

must be used for the production of hydrogen by the electrolysis at the nearest coast to the desert. The electrolysis of seawater gives us a new problem. A large amount of H_2 is required for the conversion of CO_2 emitted in the whole world to CH_4. However, a large amount of chlorine release by reaction (7.5) is not allowed for the production of H_2. Therefore, the electrode which is capable of evolving oxygen without chlorine evolution must be tailored. Oxygen evolution is the reverse reaction of reaction (7.3).

$$4\,OH^- = O_2 + 2\,H_2O + 4\,e. \tag{7.3a}$$

Investigations in tailoring amorphous-alloy anodes and cathodes for the electrolysis of seawater are in progress.

Because of large-scale production of CO_2, the catalytic reaction for CO_2 conversion must be very fast under ambient pressure on a catalyst without containing any platinum-group element. CH_4 formation is the only possible reaction to be used, since reactions to form other hydrocarbons are too slow to be used in addition to usually requiring high pressure. Amorphous alloys are very effective catalyst precursors for the conversion of CO_2 by the reaction (7.8).

$$CO_2 + 4\,H_2 = CH_4 + 2\,H_2O. \tag{7.8}$$

Figure 7.21 shows the rate of CO_2 conversion to form CH_4 on various catalysts prepared from amorphous alloys. In the atmosphere of reaction (7.8), valve metals are oxidized, but nickel is in the metallic state. Therefore, these amorphous alloys are at first oxidized in air and then reduced in a hydrogen atmosphere. The catalyst thus formed consists of fine grains of valve-metal oxides on which nickel in the metallic state is finely dispersed. The single-phase nature of amorphous alloys is very suitable for forming this kind of catalyst which cannot be prepared by any other methods.

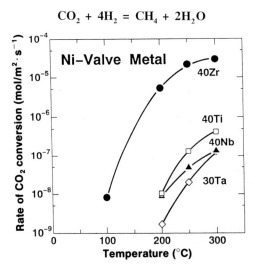

Fig. 7.21. Rates of CO_2 conversion on oxidation–reduction treated Ni-valve metal alloys as a function of reaction temperature

Since effective anode and cathode for electrolysiss of seawater and catalysts for conversion of CO_2 into CH_4 have been tailored, the global CO_2 recycling plant has been built on the roof of the Institute for Materials Research, Tohoku University [7.21]: the plant consists of solar cells for power generation at a desert, an electrolytic cell for H_2 production and a reactor for conversion of CO_2 into CH_4 at a coast closest to the desert, and a CH_4 combustion and CO_2 recovery plant at an energy consuming district. The performance of the global CO_2 recycling plant utilizing these novel electrodes and catalyst has substantiated that the solar energy at deserts can be used in the form of CH_4 without emitting CO_2 at an energy consuming district such as Japan.

7.4.2 Catalysts for the Decomposition of NO_x

NO_x is formed at temperatures higher than 1000 °C but is unstable at lower temperatures. Accordingly, the direct decomposition of NO_x into O_2 and N_2 is thermodynamically feasible at lower temperatures, such as

$$2 NO = N_2 + O_2. \tag{7.9}$$

Some amorphous iron-group metal-valve metal alloys containing a very small amount of platinum-group elements were used for the direct decomposition of NO. When the alloys were used as the heterogeneous catalyst for the decomposition of NO, oxidation of the alloys initially occurred, releasing N_2. After almost complete oxidation of the alloy specimens, direct decomposition of NO into O_2 and N_2 took place. The activities for the direct decomposition are illustrated in Fig. 7.22. The catalyst prepared from the palladium-containing alloy has the highest catalytic activity for NO decomposition which was higher than that of palladium black powder. When amorphous Ni–Ta and Ni–Nb alloys containing a small amount of palladium were used, the decomposition

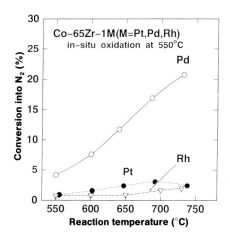

Fig. 7.22. Conversion of NO into N_2 after oxidation of Co–65 Zr–1 wt.% platinum-group element alloy specimens after in-situ oxidation at 550 °C; 0.5 vol.% NO in He was passed through 0.08 g of the alloys fixed in a length of about 40 mm in a quartz tube of 8 mm inner diameter

rate was further increased. In particular, Ni–Ta–Pd alloys showed a high activity in the temperature range from 500 to 900 °C. This high activity in the wide temperature range has never been shown by other known catalysts. It was found that the catalysts consisted of a double oxide of nickel and valve metal on which metallic palladium was finely dispersed [7.22].

7.4.3 Catalysts for the Decomposition of Chlorofluorocarbons

A leakage of stored ChloroFluoroCarbons (CFCs) leads to the destruction of the ozone layer. Accordingly, the CFCs produced so far must be converted into safe and usable substances. Various procedures have been proposed. First of all, CFCs must be decomposed not into chlorine-containing organic compounds but completely into inorganic substances. The simplest, easiest and cheapest method for the complete decomposition of CFCs is catalytic hydrolysis, i.e., the reaction of CFCs with H_2O to form CO_2, HCl and HF. This type of reaction is exothermic. For example CFC-12 decomposes as:

$$CCl_2F_2 + 2H_2O = CO_2 + 2HCl + 2HF. \tag{7.10}$$

When humid CFC-12 was passed through amorphous iron-group metal-valve metal alloys, catalytic decomposition occurred at a high rate. Amorphous Fe–Zr and Ni–Zr alloys were particularly effective as the precursor of the catalyst, although they suffered oxyhalogenation to form a mixture of oxides, chlorides, fluoride, oxychlorides and oxyfluorides. When chromium, molybdenum and/or tungsten, which have a higher resistance against oxychlorination than iron and nickel do, were added to the alloys, the durability of the catalyst was significantly increased.

7.5 Concluding Remarks

Various unknown, attractive and useful properties of amorphous alloys have been found. As has been shown already, the formation of a single-phase solid solution supersaturated with various alloying elements exceeding the solubility limits at equilibrium is quite suitable in obtaining novel materials with specific functions. The global atmospheric and energy problems can be solved by tailoring new materials by using unique properties of amorphous alloys such as catalytic, electrocatalytic and corrosion-resistant properties as well as the metastability itself. Studies of the chemical properties of amorphous alloys have not been widely performed. Further investigations will open a new era for chemical applications of novel amorphous alloys.

References

7.1 M. Naka, K. Hashimoto, T. Masumoto: J. Jpn. Inst. Metals **38**, 835 (1974)
7.2 K. Hashimoto: In *Passivity of Metals and Semiconductors*, ed. by M. Froment (Elsevier, Amsterdam 1983) p. 235
7.3 R.M. Latanision, A. Saito, R. Sandenbergh, S.-X. Zhang: In *Chemistry and Physics of Rapidly Solidified Materials*, ed. by B.J. Berkowitz, R.O. Scattergood (Metallurgical Soc. of AIME, New York 1983) p. 153
7.4 R.B. Diegle: J. Non-Cryst. Solids **61/62**, 601 (1984)
7.5 M.D. Archer, C.C. Corke, B.H. Harji: Electrochim. Acta **32**, 13 (1987)
7.6 K. Hashimoto, T. Masumoto, S. Shimodaira: In *Passitivity and Its Breakdown on Iron and Iron Base Alloys*, ed. by R.W. Staehle, H. Okada (NACE, Houston, TX 1975) p. 34
7.7 D. Huerta, K.E. Heusler: Proc. 9th Int'l Congr. Metallic Corrosion (Nat'l Research Coucil of Canada, Ottawa 1984) Vol. 1, p. 222
7.8 K. Hashimoto, M. Naka, K. Asami, T. Masumoto: Corros. Sci. **19**, 165 (1979)
7.9 H. Habazaki, S.-Q. Ding., A. Kawashima, K. Asami, K. Hashimoto, A. Inoue, T. Masumoto: Corros. Sci. **29**, 1319 (1989)
7.10 B.-P. Zhang, H. Habazaki, A. Kawashima, K. Asami, K. Hashimoto: Corros. Sci. **33**, 103 (1992)
7.11 K. Hashimoto, N. Kumagai, H. Yoshioka, J.-H. Kim, E. Akiyama, H. Habazaki, S. Mrowec, A. Kawashima, K. Asami: Corros. Sci. **35**, 363 (1993)
7.12 K. Hashimoto, P.-Y. Park, J.-H. Kim, H. Yoshioka, E. Akiyama, H. Habazaki, A. Kawashima, K. Asami, Z. Grzesik, S. Mrowec: Mater. Sci. Eng. **A198**, 1 (1995)
7.13 P.-Y. Park, E. Akiyama, A. Kawashima, K. Asami, K. Hashimoto: Corros. Sci. **38**, 397 (1996)
7.14 H. Habazaki, J. Dabek, K. Hashimoto, S. Mrowec, M. Danielewski: Corros. Sci. **34**, 183 (1993)
7.15 H. Mitsui, H. Habazaki, K. Asami, K. Hashimoto, S. Mrowec: Corros. Sci. **38** (1996) 1431
7.16 N. Kumagai, A. Kawashima, K. Asami, K. Hashimoto: J. Appl. Electrochem. **16**, 565 (1986)
7.17 N. Kumagai, Y. Samata, A. Kawashima, K. Asami, K. Hashimoto: J. Appl. Electrochem. **17**, 347 (1987)
7.18 K. Hashimoto, N. Kumagai, H. Yoshioka, K. Asami: Mater. Manuf. Processes **5**, 567 (1990)
7.19 K. Hashimoto: Mater. Sci. Eng. A **179/180**, 27 (1994)
7.20 K. Hashimoto: Trans. Mater. Res. Soc. Jpn. A **18**, 35 (1994)
7.21 K. Hashimoto, E. Akiyama, H. Habazaki, A. Kawashima, K. Shimamura, M. Komori, N. Kumagai: Zairyo-to-Kankyo (Corrosion Engineering), **45**, 614 (1996)
7.22 M. Komori, E. Akiyama, H. Habazaki, A. Kawashima, K. Asami, K. Hashimoto: Appl. Catal. B., Environmental **9**, 93 (1996)

8 Selected Examples of Applications

H.H. Liebermann, N.J. Grant, and T. Ando

Rapid Solidification Processing (RSP) has been applied to a wide range of alloys, and in limited cases to ceramics, to improve mechanical, electrochemical and physical properties for structural and non-structural applications. By RSP it is possible to produce metastable microstructures in materials which are not obtainable by conventional processing. The metastable structures in as-solidified materials are often useful for direct use, if the structures are sufficiently stable at the service temperature. Excellent examples of as-solidified magnetic and brazing alloys are described later in this chapter. As-solidified materials, however, are usually in particulate forms and are not immediately ready for most industrial applications, requiring subsequent consolidation to produce more useful, bulk forms. Additional thermal and/or thermomechanical processing may also be necessary to further adjust the microstructure for final use. Such secondary processing requires exposing the RSP material to elevated temperatures to effect diffusional processes required for structural changes in the material. The high metastability in as-solidified materials directly translates into a driving force for such structural changes. By applying appropriate secondary processing, we can control both the path and extent of the structural changes to produce a microstructure appropriate for a specific application.

In order to understand how we can exploit the opportunity for microstructural control and property enhancement, it is instructive to discuss what RSP does to the microstructure and hence to the properties of materials in generic terms. Rapid solidification has five major types of effects on material structure, namely, size-refinement, extended solid solubility, increased chemical homogeneity, evolution of non-equilibrium crystalline phases and formation of amorphous phases. Table 8.1 summarizes what property improvements we might expect to see for each of the five effects of RSP. The first three effects are important mainly for improved mechanical properties and corrosion/oxidation resistance, whereas the other two effects lead to the improvement of both mechanical and non-mechanical properties. In the next section, we will discuss how the size-refinement, extended solid solubility and improved chemical homogeneity due to RSP can be exploited to enhance the mechanical properties and the resistance to environmental degradation. The discussions are mainly on the general approaches to applying RSP, although some specific applications are also indicated where appropriate. These fundamental approaches apply to ceramic materials as well. Applications to magnetic materials and brazing alloys

218   8 Selected Examples of Applications

Table 8.1. Key structure-property relationships in RSP materials

Structural features	Expected property improvement
Size-refinement	Hall-Petch strengthening
	Improved fracture and impact toughness
	Enhanced superplasticity
Extended solid solubility	Increased solid solution strengthening
	Increased precipitation and dispersion strengthening
	Minimized brittle equilibrium phase precipitation
Chemical homogeneity	Improved corrosion/oxidation resistance
	Better response to working
	Better response to heat treatments
Precipitation of non-equilibrium crystalline phases	Improved physical properties
	Improved mechanical properties
Formation of amorphous phases	Improved physical properties
	Improved mechanical properties (strength, wear)
	Improved corrosion resistance

are discussed in relation to specific applications in the sections that follow. The final section of this chapter addresses current limitations and some future directions of the application of RSP to materials processing and manufacturing.

8.1 Improvement of Mechanical Properties

8.1.1 Size Refinement

Size-refinement refers to the production of a polycrystalline microstructure with refined sizes of matrix grains, and of any other second phases, in an RSP material. The size-refinement in RSP is thermodynamically possible because of a large chemical driving force available for the transformation of the parent liquid phase to the product crystalline phase(s). A thermodynamic calculation based on an assumed lowest nucleation temperature indicates that the smallest grain size achievable by rapid solidification of a liquid phase is of the order of 10 nm [8.1]. In this calculation, it is assumed that all of the driving force available at the nucleation temperature is converted to the grain boundary energy. In reality, RSP usually results in grain sizes greater than the theoretical limit of 10 nm because the actual nucleation temperature is likely to be higher than the assumed limit. Growth and coarsening of crystalline grains during solidification and subsequent secondary processing also give rise to an increase in grain size. Finer grains, however, can be produced in a controlled manner by crystallizing an amorphous phase.

Actual crystalline grains in a rapidly solidified alloy are often dendritic. The dendritic morphology is lost during subsequent consolidation and other thermomechanical processing which produces more equiaxed, spherical grains through recrystallization and other phase transformations. The resultant spherical grains typically show a grain size ranging from about 0.1 to 10 µm, depending on the alloy being processed and the types of primary RSP and secondary processing used. Although these grain sizes are much larger than the theoretical limit, they are still in the range exploitable for improving material properties. An important property that can be improved by size-refinement, regardless of alloys, is strength. The strengthening comes from the Hall-Petch effect, which is expressed by the relation $\sigma = \sigma_0 + kd^{-1/2}$, where is σ the yield strength, σ_0 and k are constants characteristic of the alloy of interest and d is the grain diameter. For a plain-carbon low-carbon steel which has σ_0 and k values of, respectively, 190 MPa and 17 MPa mm$^{1/2}$, a decrease in grain size from a conventional value of 30 µm to 5 µm increases the yield strength from about 290 to 430 MPa. If we can produce a 1 µm grain size by an appropriate RSP method, the same steel would show a yield strength exceeding 700 MPa, a strength value comparable to those of hardened and tempered medium-carbon alloy steels. Such a fine-grained steel may be used for critical structural parts in replacement of heat-treated medium-carbon alloy steels.

Such grain-size strengthening has been confirmed in a 0.2 wt.% carbon plain-carbon steel (AISI 1020) by the combined processing of spray forming and hot rolling [8.2]. Spray forming is an RSP method in which a molten spray of an alloy, usually produced by gas atomization, is deposited onto a substrate to consolidate the molten spray into bulk form. In this example, an AISI 1020 steel alloyed with a low concentration of aluminum, was atomized with nitrogen gas and deposited onto a cooled substrate. After rolling at a low austenitic temperature, the spray-formed steels showed a uniform ferritic grain size of about 2.5 µm as shown in Fig. 8.1. The fine ferritic grains resulted from fine prior austenite grains which were pinned by fine particles of AlN that formed during hot rolling. This steel showed a yield strength of about 600 MPa. Figure 8.2 plots the yield strength values of spray-formed 1020 steels (designated LDC) with and without aluminum additions against $d^{-1/2}$ together with those of similar low carbon steels produced by conventional processing. A Hall-Petch behavior is apparent, confirming that the observed strengthening is due mainly to the grain size effect.

Similar methods can be used to obtain fine-grained microstructures in other ferrous and non-ferrous alloys. Potential areas include high-strength low-alloy aluminum for room-temperature applications. Alternatively, higher quench-rate RSP methods to produce rapidly solidified particulate and consolidate it into bulk form by an appropriate method (such as hot isostatic pressing, hot extrusion, hot pressing and explosive compaction) may also be used where the increased cost is justified. The highest degree of grain refinement is achieved by the devitrification of an amorphous alloy, the resultant grains being as small as 10 nm.

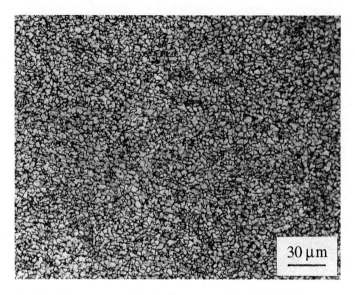

Fig. 8.1. Microstructure of an aluminum-microalloyed AISI 1020 steel processed by spray forming and hot rolling

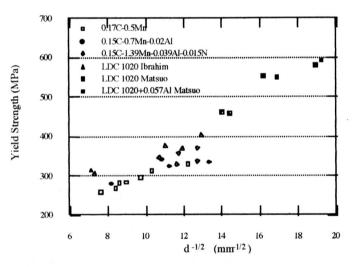

Fig. 8.2. Hall-Petch plot of low-carbon, plain carbon steels. The high strength values were achieved in aluminum-microalloyed AISI 1020 steels processed by spray forming and hot rolling

8.1 Improvement of Mechanical Properties

Size-refinement by RSP leads to improved toughness as well as high strength, irrespective of alloys. In fact, grain refinement is the only way by which both strength and toughness can be improved simultaneously and thus usually is a preferred approach to the microstructural control for improved mechanical properties. RSP also assures size refinement of all other features in the microstructure, including second phases that precipitate during solidification. Carbide phases in tool steels, for example, precipitate on much finer scale when RSP is applied [8.3]. This improves the toughness and also the response to hot working and heat treatments of tool steels. High-performance tool steels, particularly high-speed tool steels, are produced by gas atomization, consolidation by hot isostatic pressing (HIP) and additional hot working. These high-speed tool steels are referred to as power metallurgy (PM) high-speed tool steels.

Fine carbide sizes in tool steels are preferred also to improve their grindability after hardening heat treatments. For example, conventionally processed high-speed tool steels contain only up to 2 wt.% vanadium because higher vanaduim contents result in excessive precipitation of coarse vanadium carbide making the grindability of these steels unacceptably poor. Due to the size-refinement effect, higher vanadium contents exceeding 4 wt.% are permitted in PM high-speed tool steels to improve tool life without impairing grindability.

Unwanted inclusions which inevitably exist in commercial alloys in blocky, irregular forms pose much less problems if they are refined by RSP into fine, uniformly scattered particles. This would mean that alloys for critical applications can be produced from scrap which contains impurities, leading to significantly reduced alloy production costs. Examples of such critical alloys include high-strength aluminum alloys for aircraft applications. A preliminary study with a spray-formed 7150 alloy with deliberate additions of iron and silicon (which are the most commonly found impurities in aluminum scrap) indicated no appreciable negative effects of the high impurity levels studied. This is by no means a trivial area of RSP application, particularly in view of the enormously large energy savings as well as cost reduction achievable by the substitution of scrap for low-impurity, energy-intensive virgin aluminum in the alloy production.

Another important property to which size-refinement may lead is superplasticity. For superplaticity to occur in an alloy, the alloy must contain fine, equiaxed grains smaller than about 10 µm so that deformation can occur predominantly by grain boundary sliding. Superplastic deformation at industrially useful high strain-rates requires even smaller grain sizes, often near, or preferably less than, 1 µm. Such fine grains are difficult to produce economically. They are also hardly stable by themselves and coarsen very quickly at the deformation temperature unless a second phase is used to increase the stability. Such a second phase, too, must be in a fine, uniform distribution to assure a uniform stability of the microstructure. Needless to say, RSP is an excellent method for producing such fine, stable microstructures and has been applied to many alloys to enhance superplasticity in them [8.4–6].

Superplastic forming (SPF) is suitable for manufacturing complex shapes

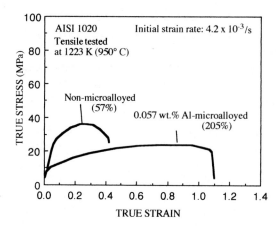

Fig. 8.3. Tensile stress-strain curves at 950 °C for spray formed and hot rolled AISI 1020 steels. Microalloying with aluminum stabilizes the microstructure and hence increases ductility

and has been applied to many aircraft structural components. Many RSP nickel-based and cobalt-based superalloys, titanium alloys and aluminum alloys are developed for the SPF applications. RSP has also been applied to enhance superplasticity in alloys containing a large content of hard phases such as untrahigh-carbon steels and Fe–C–B–Cr–Ni alloys. The application of RSP for enhanced superplasticity is expected to increase as more economical RSP methods are developed. One of the RSP methods which has such potential is spray forming. The AlN-stabilized fine-grained 1020 steel shown in Fig. 8.1, for example, showed an elongation of 205% at an austenitic temperature of 950 °C and a very low flow stress of about 10 to 20 MPa as shown in Fig. 8.3. The same steel without grain-boundary pinning by AlN particles showed only a 57% elongation and a much higher flow stress.

8.1.2 Extended Solid Solubility

Increased solid solubilities beyond equilibrium values are often observed in RSP alloys. Thermodynamically, a range of solid compositions is permissible at the solid/liquid interface for a given interfacial liquid composition if the interface is kept below the equilibrium liquidus temperature [8.1], as is the case in rapid solidification. At high interface velocities, the interfacial solute partition coefficient, k, defined by the ratio of the solute concentration in the solid to that of the liquid, deviates from its equilibrium value, and if the interface is undercooled sufficiently, approaches unity at extreme interface velocities exceeding the diffusional velocities of the solute elements. The resultant solid phase thus contains more solute (if $k < 1$) than permitted by equilibrium thermodynamics. This phenomenon, called solute trapping, gives rise to extended solid solubility in an RSP alloy.

The extended solid solubility has two major effects on the resultant microstructure. First, it assures reduced precipitation of coarse second phases which are detrimental to mechanical properties. Good examples of such effects are found in RSP high-strength aluminum alloys (e.g., 7XXX, 5XXX and 2XXX alloy series) in which coarse, brittle intermetallic phases are eliminated. The brittle intermetallic phases have complex crystal structures and thus have little chance to precipitate during the short time available in rapid solidification [8.7]. Other simpler equilibrium second phases may be permitted to precipitate. Metastable phases may also precipitate in replacement of the equilibrium phases. The usual RSP methods applied to aluminum alloys are gas atomization and melt spinning followed by consolidation by hot extrusion of canned powder. Melt-spun alloys require mechanical comminution before consolidation. Hot extrusion is necessary for RSP aluminum alloy particulate to break oxide film which exists on the particulate surface by the large attendant shear deformation. The resultant microstructure after consolidation presents no complex intermetallic phases and is characterized by fine, uniform matrix grains and a high degree of chemical homogeneity, giving rise to improved strength, toughness and stress-corrosion cracking resistance. An alternative RSP method applicable to aluminum alloys is spray forming which leads to lower costs, less contamination or oxidation than the particulate approach, and often improved properties, although the range of alloying permitted in spray-formed alloys may be somewhat limited [8.8].

The improved strength of RSP high-strength aluminum alloys comes partially from the refined grain sizes (Hall-Petch effect) but more importantly from increased precipitation strengthening. An increased solid solubility in an RSP alloy can also be used to increase dispersion strengthening for elevated-temperature applications. This is especially advantageous in aluminum alloys since aluminum has very low equilibrium solid solubilities for many alloying elements. Aluminum alloys containing high contents of transition metals, such as Fe, Mo, Cr, Ni, V and Ce, have been processed by RSP for this purpose [8.7]. The major alloy systems studied include Al–8 Fe–2 Mo, Al–4 Cr–3 Zr, Al–5 Fe–3 Ni–6 Co, Al–12–2 V and Al–8 Fe–4 Ce. Although the equilibrium solubilities of these elements are low, RSP leads to high degrees of supersaturation as well as refinement or elimination of second phases and, upon subsequent aging, permits controlled precipitation of useful second phases at a high concentration to effect dispersion strengthening in these alloys. The second phases that precipitate in these high-temperature alloys are simple stoichiometric compounds between aluminum and the transition metals. These second phases are often in metastable forms but are stable enough to sustain as useful dispersoids at the service temperature. With these RSP aluminum alloys, high strength matching the density-compensated strength of titanium alloys has been achieved at temperatures up to 250 °C. Applications of such RSP high-temperature aluminum alloys include airframe, missile and engine parts.

When applied to aluminum–lithium alloys, RSP permits increased additions of lithium up to about 4 wt.%, whereas conventional ingot metallurgy can tolerate less than 2 wt.% because of the precipitation of brittle intermetallic phases that occurs during slow solidification. Increases in lithium contents in

RSP alloys lead to further increases in elastic modulus and strength and to as much as a 14% decrease in density. The increased modulus and strength come from the precipitation of the coherent intermetallic phase (Al_3Li) upon aging [8.9]. Another alloy system that is important for high-modulus low-density applications is the Al–Be system. High Be contents exceeding 10 wt.% have been used to decrease the alloy density as much as 25% at an excellent strength level. Ternary Al–Li–Be alloys have also been studied.

Increased precipitation and dispersion strengthening are achieved in other RSP alloys including magnesium, nickel and copper alloys. Copper alloys are used in applications where their high electrical or thermal conductivity is useful. Their use, however, is often limited due to their insufficient strength, particularly at elevated temperatures. Solid-solution strengthening and precipitation strengthening usually degrade the conductive properties and thus are not preferred approaches. Improved strength has been achieved without significant loss of conductive properties in RSP copper alloys with increased contents of transition metals [8.10]. The strengthening is due to fine dispersions of stable intermetallic phases which are insoluble in the copper matrix. Such RSP copper alloys find applications in critical components such as rocket thrust chambers where high thermal conductivity and strength are both required. RSP has enabled modifying the compositions of Ni-based superalloys for increased precipitation and/or dispersion strengthening by γ'-phase and stable carbides. RSP magnesium alloys have been studied for increased specific strength and elevated-temperature strength [8.11]. Many of these RSP alloys have been applied to critical aerospace and automotive components for which improvement of more than one property is necessary.

8.1.3 Chemical Homogeneity

The limited time available for diffusional processes during rapid solidification gives rise to minimum segregation of alloying elements in RSP alloys. Structurally, this is manifested by a decrease in phase separation as well as in compositional variations within a phase.

Because of the chemical homogeneity in the microstructure, RSP alloys generally show excellent response to thermal and thermomechanical processing. For example, PM high-speed tool steels show a higher incipient melting point than cast counterparts. This makes hot working more practical for a wider range of alloy compositions. Hardening of these tool steels can also be done under more relaxed austenitizing conditions without the fear of partial melting or uneven structural coarsening. Such an increase in incipient melting temperature by RSP also permits alloy modifications in nickel-based superalloys to increase the γ' solvus temperature, leading to increased γ' precipitation and hence to improved high-temperature strength in these alloys. Heat treating RSP alloys is generally easier; consolidated RSP alloys require little or no homogenization treatments because of short diffusion distances in RSP alloys. Solution treatments of RSP alloys require shorter times than necessary for conventional alloys. Choosing a precise solutionizing temperature is less critical for RSP alloys in which fine grains are stabilized. Austenitizing of steels with particle-pinned austenitic grain boundaries (Fig. 8.1) can be done even outside the austenitizing temperature ranges used for conventionally processed steels.

The RSP-enhanced chemical homogeneity improves the resistance to environmental degradation. RSP aluminum alloys, for example, show less tendency for pitting corrosion in salt-fog environments. This is attributed to a reduced distance between adjacent pitting sites so that local galvanic cells shield one another. Cu–Mn alloys have a good acoustic damping capacity due to a magnetoelastic interaction effect, but their use in corrosive environments are limited because of their susceptibility to Stress-Corrosion Cracking (SCC) and low fracture toughness. RSP minimizes alloy segregation in these alloys and hence improves SCC resistance and fracture toughness. RSP has also been applied to Cu–Ni–X alloys, where X is a transition metal, to improve both improved seawater-corrosion resistance and strength at once. Amorphous alloys generally have high intrinsic corrosion resistance because of their ultimate compositional uniformity.

RSP improves the oxidation resistance of high-temperature alloys such as stainless steels and superalloys. The proposed mechanisms are based on an increased tendency of protective scale formation [8.12]. In RSP stainless steels with a fine grain size, this is achieved by the fast diffusion of chromium through grain boundaries to the surface where a chromium oxide protective layer quickly forms. Without a stable fine grain size, lattice diffusion dominates, making the surface deficient in chromium. The alternative oxide that forms is an iron oxide which does not protect the steel from further oxidation.

8.2 Magnetic Applications

The potentially attractive basic magnetic properties of rapidly solidified alloys date back prior to 1970. The application of such alloys, though, required the ability to produce a geometrically regular form of these materials, such as sheet. The first rapidly solidified alloys to be produced in such a form were made by AlliedSignal Amorphous Metals, then Allied Chemical. The process of planar-flow casting allowed the production of geometrically uniform sheet of rapidly solidified magnetic and other alloys. Of the rapidly solidified magnetic alloys, amorphous alloys have found substantial applications in areas such as electrical power distribution, high-frequency magnetics, electronic article surveillance systems, and others. Rapidly solidified crystalline (microcrystalline) alloys have found applications as high-efficiency transformer-core steels and very high-energy product permanent magnet applications.

8.2.1 Magnetic Properties and Applications

Soft magnetic materials are, by definition, easily magnetized and demagnetized by the application of a magnetic field. These materials are typically heat treated in a magnetic field to reduce impediments to magnetic domain wall motion and thus reversal of magnetization. Conventional soft magnetic materials are also mechanically soft, as shown in Fig. 8.4. Amorphous alloys, however, are an exception to this rule, exhibiting a unique combination of excellent soft magnetic properties and high mechanical strength. Magnetically hard materials are usually very hard mechanically as well. These materials are used in various applications requiring a source for a magnetic field which is unperturbed by heat, extraneous magnetic fields, and other environmental conditions. Very high energy product permanent magnets based on Nd–Fe–B alloy have been developed and commercialized. Applications include their use in the numerous electric motors which are found in automobiles.

The use of rapidly solidified alloys in various applications depends on the specific demands of those applications. For example, applications involving the handling of large quantities of electrical power will require high saturation induction of the alloy. Other applications, such as high-frequency transformers, require lowest losses. Still other applications, such as saturable-core reactors and magnetic amplifiers, require high permeabilities. The point is that specific magnetic properties of rapidly solidified alloys need to be developed by alloy design and/or by subsequent heat treatment to meet the demands of the application at hand. As it turns out, it is usually a *combination* of properties that make a material suited for a particular application. For example, magnetic tape heads

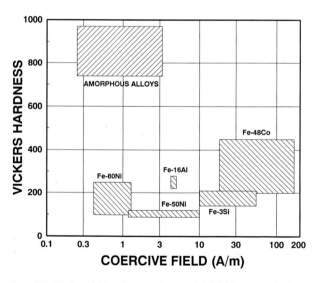

Fig. 8.4. Mechanical hardness vs dc coercive field for several soft magnetic materials

require the magnetic material to have both high permeability at high frequencies and also wear resistance. Sensor applications, on the other hand, rely on properties such as Young's modulus, anisotropy, and magnetostriction to provide the required changes in magnetic properties with stresses imposed upon the sensing element.

8.2.2 Power Magnetic Applications

The transmission and distribution of electrical power in this and other countries is accomplished through the use of a power grid in which line-frequency transformers are used to step down voltages to residential and commercial customers. The losses incurred through the use of such a system is enormous. The use of amorphous magnetic alloys instead of the conventionally used silicon–iron alloys represents tremendous potential cost savings in the form of wasted energy saved. Distribution transformers at the final step of the power grid, and their number and potential cost savings are also tremendous. While a drawback of amorphous alloys used in transformers is their lower saturation induction, in comparison to that of silicon steels, the core losses at operating inductions is only about one fourth of that of the conventional silicon irons used today. As with conventional transformer cores, those made of amorphous alloys also need to be heat treated in a magnetic field to optimize performance characteristics.

The final step in the electrical power-distribution network is the distribution transformer, which also has the greatest installed capacity. Therefore, an excellent opportunity exists for installing a large number of high-efficiency amorphous alloy-cored transformers each year. The concept of loss evaluation by the utilities results in establishing the value of using amorphous alloy-cored distribution transformers. While the purchase price of an amorphous alloy-cored transformer is greater than that a conventional transformer, the total cost over its operation life is significantly less. Various wound-core and stacked core transformer designs have been implemented over the years. Experience with amorphous alloy-cored transformers in the field has been very good so far.

8.2.3 Specialty Magnetic Applications

There are numerous applications for soft magnetic materials in the form of small cores for electronic devices. The characteristics of amorphous alloys in these applications include high electrical resistivity to result in reduced eddy current losses and high Curie temperature.

An application in which inductive components are used is in switched-mode power supplies, which maintain a regulated output voltage by using feedback to change the duty cycle of a high-frequency switch between source and load. The same factors which make amorphous alloys attractive for used in switched-mode power supplies also apply to transformers used in induction heating. Other specialty magnetic components such as output chokes and magnetic amplifiers, and EMI filters are all well addressed by the capabilities of rapidly solidified alloys after suitable heat treatment.

Transducers convert energy from one form to another and are based on various physical phenomena. Many transducers provide a voltage output to indicate force, pressure, temperature, position, etc. On the other hand, some transducers cause various physical actions, based on an input voltage. This class of transducers include devices such as loudspeakers, ultrasonic vibrators, and positioners. The high permeability and good frequency response of ferromagnetic amorphous alloys make them good candidates for replacing materials presently used in these applications.

Sensors used in the entertainment include magnetic tape heads, dynamic microphones, and phonograph cartridges. Sensors used for measurements cover a wide range of properties, the most widespread of which is the measurement of magnetic field. In this case, the sensing capabilities of the magnetic material alone are essential to the operation of a device. Other sensors utilize the rapidly solidified alloy magnetic properties in the form of device components, such as speed, proximity, and position sensors. A large variety of sensors can be designed by using the magnetomechanical sensitivity of amorphous alloys in combination with their robust mechanical properties. Such sensors measure strain, torque, and even sound velocity.

A unique application of rapidly solidified alloys is in the area of Electronic Article Surveillance (EAS), in which tags attached to merchandise to be controlled are detected if they pass through an access gate. The basic types of magnetic EAS tags are two: resonant and harmonic. In the resonant tags, the magnetic element is caused to resonate mechanically when excited by a high-frequency magnetic field and the damping of this excitement is then monitored. In the resonant tags, the magnetic element is again excited with a high-frequency magnetic field and a harmonic is monitored as signature of the tag. High harmonic content of such tags requires a very square magnetization loop.

Because of the very high magnetic permeability of suitably annealed amorphous alloys, and because of their concurrent mechanical strength, their use in magnetic shielding applications has also developed. Specific shielding applications include those to exclude EMI interference, magnetostatic shielding for superconducting magnets, and applications such as magnetically shielded floppy disk boxes and even magnetically shielding venetian blinds.

Applications involving pulsed power for accelerators and lasers are a relatively new development. Amorphous alloys are ideally suited for these kinds of applications because of their intrinsically high resistivities, thin gauge, and very high strength. In addition, the amorphous alloys having high saturation induction allow reasonably compact devices to be constructed. Specific applications of amorphous alloys in pulsed-power include devices such as pulse transformers, switch protection inductors, and magnetic pulse inductors.

8.3 Joining Applications

Materials joining by brazing and soldering is an ancient practice that has evolved into a sophisticated modern technology through various innovations, such as the use of rapidly solidified filler metals. Much of conventional joining technology can be utilized when considering the use of rapidly solidified filler metal, while substantial benefits in component reliability and performance is realized. In many applications, a thin layer of filler metal is placed between the two base materials to be joined. Heat is then applied to the entire pre-placed "sandwich" and melting of the filler metal ensues. The entire assembly is then allowed to cool and the filler metal forms an integral, strong bond between the base metal pieces. Specific examples of commercial components which are brazed using amorphous alloy filler metal are shown in Fig 8.5. The design and brazing of honeycomb structures is depicted in Fig. 8.6.

The advantages of rapidly solidified alloys are many: ductility, homogeneity of microstructure, low melting temperature, thin gauge, and the possibility of having self-fluxing. Many conventional brazing alloys are brittle because of the content of melting temperature reducing elements. such as phosphorus, boron, and silicon. Thus, a conventional form of brazing alloys is paste, in which a powdered form of the alloys is carried by an organic binder. Application of this form is convenient in that the filler metal composite can easily be directed at the contact point of the two base metal pieces to be joined. The drawback, however, lies in the residue and porosity which can result when using filler metal in this form. In contrast, rapid solidification can result in the production of the same filler-metal composition, but in ductile ribbon or powder form. The reason for this is that the very same elements which are added to the filler metal to depress the melting point are also glass-formers for the alloy when rapidly solidified. Thus, a thin, ductile, low-melting-temperature filler metal results. In addition, ductile foil can be stamped into preforms, allowing the gap between the base metal pieces to be more completely and efficiently filled during joining. In fact, some phosphorus-containing amorphous alloy filler metals can be used to braze copper without the use of a cleaning flux. While the filler metal in a joint is no longer amorphous after brazing, its microstructure is extremely fine, and is directly associated with the fineness of microstructure of the starting filler metal ribbon.

The use of various solders has also benefited by the use of rapidly solidified alloy. While not amorphous on rapid solidification, solders show a refined crystalline microstructure in a single step production operation. The refined microstructure naturally lends itself to improved joining operations in automated die bonding, for example, in which silicon chips atop solder-foil chips on a leadframe are continuously run through a tunnel oven. As with brazing alloys, the ductility of many solders can be improved through the use of rapid solidification processing, thereby facilitating their use.

Applications in which rapidly solidified filler metals are used are varied. For

Fig. 8.5. Typical jet engine parts fabricated with METGLAS alloy brazing foil: *1* Exhaust plug for Pratt & Whitney JT9D-7R4 jet engine; *2* tailpipe for Douglas DC-9-50 aircraft; *3* exhaust plug for GE CF6-80 jet engine; and *4* outer nozzle ring for Rolls Royce RB-211 jet engine

example, abradable seals for aircraft engines was one of the first commercial applications of amorphous alloy filler metal. In other cases, amorphous alloy filler metal has been used to replace much more expensive conventionally used gold-based filler metal. Another application in which rapidly solidified foil-form filler metal can be advantageously used is in heat exchangers, in which intricate honeycomb structures need to be precisely and reliably joined to face plates. Because of the nature of RSP, tailoring of filler-metal alloy chemistry to suit specific needs is relatively simple, because a ductile foil-form product almost

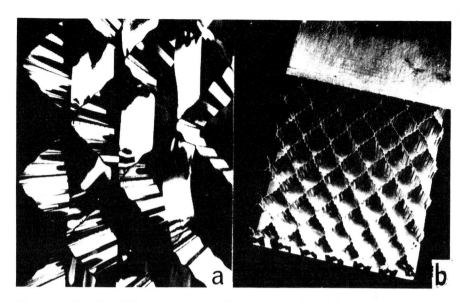

Fig. 8.6a, b. Rapidly solidified brazing foil enables the construction of closed-cell, brazed, honeycomb structures. Foil is sandwiched between the nodes of a honeycomb core (**a**), face sheets are placed on the core (**b**), and the entire assembly is then brazed in a vacuum furnace. During the brazing operation, the filler melts and flows to the face sheet/core interface, forming a uniform, continuous fillet. The structure thus produced is used for acoustic tail pipes on Boeing 727 aircraft

always results. Other applications of rapidly solidified filler metals include the joining of cemented carbide tips to bits for oil well drills, to metal/ceramic joining, and even to orthodontic devices.

8.4 Current Limitations and Future Directions

Despite expensive efforts that have been made to study and commercially apply RSP, its acceptance, particularly by mass-producing industries, has not been sufficient. The primary reason for this is the high cost. This arises from the many process steps required in RSP, particularly in particulate processing, that are needed to produce final products and the lack of mass-producing capabilities at each process step. An important process step that presents a bottle neck is the production of the rapidly solidified particulate itself. While rapid solidification per se does not necessarily require a high cooling rate, as exemplified by the solidification of a highly undercooled melt where the prior cooling rate is unimportant, most industrial applications of RSP rely on a high cooling rate to insure the desired RSP effects. Usually, higher cooling rates lead to more unusual (metastable) structures and hence to more drastic property improvements.

However, the mass producibility of particulate largely depends on the cooling rate required for a specific application and decreases with increasing required cooling rates. Thus, a rational approach to economical industrial application of RSP is to identify useful applications that do not necessarily require extremely high quench rates.

In real terms, efforts for the commercial application of RSP would be best rewarded by focusing on high-tonnage alloys such, as steels and aluminum alloys, and on high production-rate RSP processes, such as gas atomization, that still assure reasonably high cooling rates of 10^3–10^6 K/s. Many gas-atomization processes have been studied for increased cooling rates within the above limits. A common approach in all rapid-solidification gas-atomization processes is to use a configuration that permits placing the high-velocity atomizing gas jets in the very close proximity of the molten metal stream. Another common feature is the use of a high gas-to-metal flow ratio. However, because of the use of high gas-to-metal flow ratios, high cooling rates in these gas atomization procedures are generally achieved only at the expense of production (mass flow) rate. A clever approach to circumvent this dilemma may be the use of a planar geometry for the molten stream instead of the more conventional cylindrical geometry. Such a gas-atomization process is termed linear gas atomization because of the linear geometry of the atomization zone. With the linear geometry, increased powder-production rates are considered to be possible while maintaining high quench rates.

The industrial acceptance of RSP, however, will still be insufficient even when we have achieved high-rate production of RSP particulate, primarily because of the expensive secondary processing of the particulate to produce the final product. Unfortunately, cost reductions in the secondary processing are not likely. An approach to circumventing the high-cost secondary processing is the use of spray forming. As briefly mentioned earlier in Sect. 8.1, spray forming

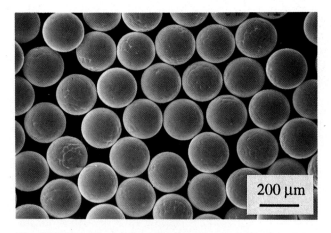

Fig. 8.7. Uniform particles of tin produced by the uniform-droplet process

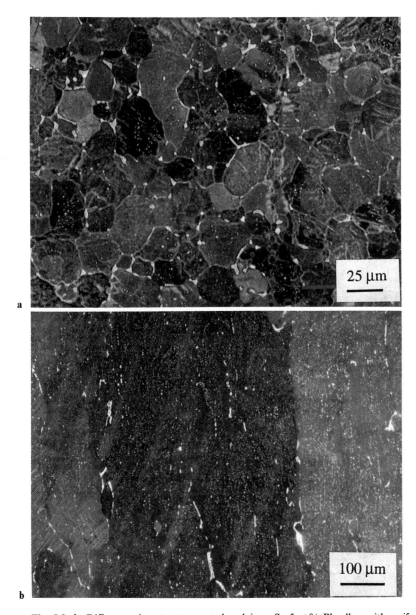

Fig. 8.8a,b. Different microstructures produced in a Sn-5 wt.% Pb alloy with uniform droplet sprays under distinctly different spray conditions: (**a**) Fine, equiaxed microstructure produced with mushy droplets and (**b**) epitaxially grown, columnar microstructure produced with molten droplets. The micrographs show transverse sections of spray deposits

produces rapidly solidified bulk alloys by the deposition of a molten alloy spray onto a substrate. Studies at the University of Swansea in Wales in the early 1970s have led to the development of many spray-forming processes. The application of spray-forming processes has been studied with many alloys, and more recently, with metal-matrix composites. Despite the relatively low solidification rates and subsequent cooling rates in spray-formed alloys, examples of improved properties are in abundance.

Another important aspect of spray forming is the ability to control the geometry of the spray-formed alloy. Simple near-net shapes, such as cylinders and disks, have been successfully produced by spray forming to replace conventional cast billets and forging stocks. The primary incentives for spray forming are to improve materials properties and to reduce processing costs relative to other RSP processes. In more aggressive efforts, spray forming has been used to manufacture products having a simple shape such as pipes and rolls. These applications are very important because they mark the earliest attempts of direct application of rapid solidification to manufacturing. The industrial acceptance of RSP would be genuinely achieved if such a direct application becomes possible in wide ranges of manufacturing.

To achieve such a goal, however, would require developing a process by which both microstructure and geometry can be controlled simultaneously at much higher levels than are currently possible. It should be stressed that, in such direct manufacturing, neither microstructure nor geometry may be altered by secondary processing, as once the geometry has been fixed there is no room for thermomechanical processing for further microstructural control. Most current spray-forming processes, however, whether commercialized or under development, are not considered to be applicable to the direct manufacturing of complex components with final microstructure, primarily because of the difficulties in achieving precise control of the solidification and motion of the droplets in the spray.

A non-gas atomization technique based on the controlled break-up of a laminar jet has been developed and studied to assess the feasibility of controlling the geometry and the microstructure simultaneously in spray forming. This process produces a spray of alloy droplets that are uniform in size. Figure 8.7 shows an SEM micrograph of a uniform tin powder produced by this method. The uniform droplets have identical thermal histories, which are dependent only on the distance from the spray nozzle, and thus permit precise control of the microstructure in the spray-formed material as shown in Fig. 8.8. Precise geometry control is expected to be achieved due to the absence of high pressure gas jets in this process.

References

8.1 J.C. Baker, J.W. Cahn: In *Solidification*, ed. by T.J. Hughel, G.F. Bolling (ASM, Metals Park, OH 1971) pp. 23–58
8.2 S.-Matsuo, T. Ando, M.C. Zody, N.J. Grant: Adv. Powder Mettallurgy **5**, 161–167 (1991)
8.3 I.R. Sare, R.W.K. Honeycombe: Met. Sci. **13**, 269–279 (May 1979)
8.4 H.N. Azari, G.S. Murty, G.S. Upadhyaya: Superplastic behavior of thermodynamically treated P/M 7091 aluminum alloy, Metall. Trans. A **25**, 2153–2160 (1994)
8.5 N.J. Grant, R.M. Pelloux: *Solidification Technology* (Battelle Labs., Columbus, OH 1974) pp. 317–336
8.6 C.P. Ashdown, Y. Zhang, N.J. Grant: Int'l J. Powder Metall. **23**, 33–37 (1987)
8.7 J.E. Hatch (ed.): *Aluminum Properties and Physical Metallurgy* (ASM, Metal Park, OH 1983)
8.8 A.I. Kahveci: In *Science and Technology of Rapid Solidification and Processing*, ed. by M.A. Otooni, NATO ASI Ser. E: Applied. Sciences, Vol. 278 (Kluwer, Dordrecht 1995) pp. 249–269
8.9 I.G. Palmer, R.E. Lewis, D.D. Crooks: In *Rapid Solidification Processing, Principles and Technologies II*, ed. by R. Mehrabian, B.H. Kear, M. Cohen (Claitor Publ., Baton Rouge 1980) pp. 324–353
8.10 I.E. Anderson, B.B. Rath: In *Rapidly Solidified Amorphous and Crystalline Alloys*, ed. by B.H. Kear, B.C. Giessen, M. Cohen (North-Holland, New York 1982) pp. 219–244
8.11 S.K. Das, C.F. Chang. In *Rapidly Solidified Amoerphous and Crystalline Alloys*, ed. by B.H. Kear, B.C. Giessen, M. Cohen (North-Holland, New York 1982) pp. 137–156
8.12 G.J. Yurek, D. Eisen, A.G. Reed: Met. Trans. A **13**, 473–485 (1982)

Further Reading

RSP Principles
Duwez P.: ASM Trans. **60**, 607–633 (1967)
Kurz W, D.J. Fisher: *Fundamentals of Solidification*, 2nd edn. (Trans Tech Publ., Adermannsdorf, Switzerland 1989)

RSP Applications
Rapidly Solidified Powder Aluminum Alloys, ed. by M.E. Fine, E.A. Starke, ASTM STM (1986)
Rapidly Solidified Amorphous and Crystaline Alloys, ed. by B.H. Kear, B.C. Giessen, M. Cohen (North-Holland, New York 1982)
Sience and Technology of Rapid Solidification and Processing, ed. by M.A. Otooni, NATO ASI Ser. E: Applied Sciences, Vol. 278 (Kluwer, Dordrecht 1995)

Recent Developments in RSP
Grant N.J.: Metall. Trans. A **23**, 1083–1093 (1992)
Cohen M.: In *Advancing Materials Research*, ed. by P.A. Psaras, H.D. Langford (Nat'l Academy Press, Washington, DC 1987)
Passow C.H., J.H. Chun, T. Ando: Metall. Trans. A **24**, 1187–1193 (1993)

RSP Principles
Aziz M.J.: J. Appl. Phys. **53**, 1158 (1982)
Baker J.C., J.W. Cahn: In *Solidification*, ed. by T.J. Hughel, G.F. Bolling (ASM, Metals Park, OH 1971)
Kurz W., D.J. Fisher: *Fundamentals of Solidification*, 2nd edn. (Trans Tech Publ., Adermannsdorf, Switzerland 1984)
Rapid Solidification Processing – Principles and Technologies II, ed. by R. Mehrabian, B.H. Kear, M. Cohen (Claitor's Publ. Div., Baton Rouge, LA 1980)

RSP Applications
H. Jones: *Rapid Solidification of Metals and Alloys* (The Inst. of Metalurgists, London 1982)
Rapidly Solidified Amorphous and Crystalline Alloys, ed. by B.H. Kear, B.C. Giessen, M. Cohen (North-Holland, Amsterdam 1982)
Rapidly Solidified Crystalline Alloys, ed. by S.K. Das, B.H. Kear, C.M. Adam (TMS, Warrendale, PA 1986)
Rapidly Solidified Powder Aluminum Alloys, ed. by M.E. Fine, E.A. Starke (Am. Soc. Testing Materials, Philadelphia, PA 1986)

Recent Developments in RSP
Amorphous Metallic Alloys, ed. by F.E. Luborsky (Butterworths, London 1983)
Cohen M.: In *Advancing Materials Research*, ed. by P.A. Psaras, H.D. Langford (National Academy Press, Washington, DC 1987)
Grant N.J.: Metall. Trans. A **23**, 1083–1093 (1992)
Passow C.H., J.H. Chun, T. Ando: Metall. Trans. A **24**, 1187–1193 (1993)
Rapidly Solidified Alloys, ed. by H.H. Liebermann (Dekker, New York 1993)
Rapidly Solidified Metals, ed. by T.R. Anantharaman, C. Suryanarayana (Trans Tech Publ., Adermannsdorf, Switzerland 1987)

Volumes in Which the Applications of Rapidly Solidified Alloys Are Specifically Discussed
Metallic Glasses, ed. by J.J. Gilman, H.J. Leamy (ASM, Metals Park, OH 1978) [papers presented at a seminar of the Materials Science Div. of the Am. Soc. for Metals]
Rapidly Quenched Metals, ed. by N.J. Grant, B.C. Giessen (Elsevier, New York 1976) [Proc. 2nd Int'l Conf. on Rapidly Quenched Metals, Cambridge, MA]
Rapidly Quenched Metals, ed. by S. Steeb, H. Warlimont (North-Holland, Amsterdam 1985) [Proc. 5th Int'l Conf. on Rapidly Quenched Metals]
Rapidly Quenched and Metastable Materials, ed. by T. Masumoto, K. Hashimoto (Elsevier, New York 1994) [Proc. 8th Int'l Conf. on Rapidly Quenched and Metastable Materials]

9 Glossary of Important Terms

T. Egami, M.A. Otooni, and W. Dmowski

Amorphization
A process by which amorphous materials are produced from vapor, liquid or solid phase by various techniques such as rapid quenching of liquid, vapor deposition, sputtering, solid state reaction, etc.

Amorphous Alloys
Metallic amorphous solids which contain multiple elements.

Amorphous Structure
Arrangement of atoms in a solid that does not form a periodic lattice as in crystalline materials. Thus, this structure has no transitional symmetry. It is not possible to define a unit cell as in crystalline solids. The diffraction pattern from the amorphous structure produced by X-ray, neutron, or electron scattering shows diffuse halos. In comparison, the diffraction pattern of crystalline material exhibits sharp spots (Bragg peaks) due to the interference of the scattered waves from the periodic atomic lattice.

Annealing
A process in which material is subjected to heating. It is frequently used to enhance desired properties, for instance, magnetic anisotropy, decrease of internal stress, or in controlled crystallization, to produce small crystalline precipitation, etc.

Atomic Pair Distribution Function
Defines a number density for atoms separated by distance **r**. Commonly used in descriptions of the atomic structure of an amorphous solid or liquid. The atomic pair distribution function can be obtained experimentally from X-ray, neutron, or electron scattering experiments by the Fourier transformation of the normalized diffraction intensity (structure factor).

Atomic Short-Range Order
Well defined (not random) arrangements of atoms in amorphous solids or liquids limited to first and second neighbors. It does not extend beyond the second atomic shell.

Atomization
A process in which a molten alloy is dispersed into small clusters by a jet of ultrasonic inert gas such as argon. Then, the atomized liquid droplets can be impacted against a cold substrate, or cooled in a water stream.

Chemical Deposition
A process to form a solid in which a chemical reaction is used. Another component is added to the bath containing the solution or other form of one element

resulting in the deposition of the desired alloy on an immersed substrate. Similar to chemical plating.

Compositional Short-Range Order (CSRO)
An atomic short-range order that involves ordering the same or different chemical elements. It is frequently observed that local arrangements of atoms in liquids or amorphous solids resemble local arrangements of crystalline materials of similar chemical composition. For instance, in $Fe_{75}B_{25}$ metallic glass, atomic CSRO is similar to that in the metastable Fe_3B crystalline solid.

Critical Cooling Rate
The minimum cooling rate that has to be achieved to produce amorphous materials. For instance, the typical cooling rate to obtain metallic glass by rapid quenching of liquid is of the order of 10^5-10^6 K/s.

Crystallization
A process that leads to the transformation of liquids or amorphous solids into crystalline phases. Conventionally, the crystallization process of amorphous solids is divided into primary, eutectic, and polymorphous crystallization. Since amorphous materials are metastable with respect to the crystalline phase, eventually they will transform into a crystalline phase. Time scale of this transformation depends on temperature. For many amorphous solids, the time of transformation is so large, even in the range of 400–500 K, that they can be safely used in technical applications.

Dense Random-Packed Hard Spheres
A model of the atomic structure of an amorphous solid. It assumes that atoms can be described as hard spheres that are densely but randomly packed. A physical model of such a structure was used in the first experimental modeling of the atomic structure before more sophisticated computer modeling was introduced. In such an experiment, steel balls were put into a bag, and after some shaking liquid wax was poured in. After the wax solidified, the coordinates of the top layer of balls were measured, and those balls were removed. Then the coordinates of the next layer of balls were determined and those balls removed, and so on to determine the positions of all the balls. Finally, numerical data were converted into a pair distribution function, which could be compared with another experiment.

Diffuse Scattering
X-ray, neutron, or electron diffraction pattern without sharp, well-defined spots (Bragg peaks). Produced by the random arrangement of atoms.

Electrodeposition
Similar to chemical deposition but electrodes and applied voltage are used in the bath to control the deposition rate. The material is deposited on an electrode.

Eutectic Crystallization
A crystallization process in which two crystalline phases are growing in a coupled fashion, analogous to eutectic crystallization of liquids. In amorphous or metastable crystalline phases a crystallization process takes place in the solid state, so the product of this reaction can be called eutectoid.

Gas Atomization
See atomization.
Glass-Formation Criteria
Necessary conditions to obtain glass from a liquid. They can specify compositional range, temperature of liquid, cooling rate etc. Frequently described in the Time–Temperature-Transformation (TTT) diagram, which assumes that glass formation is determined by the kinetics of crystallization of crystalline phases, which has to be suppressed. However, there are different theoretical approaches to the glass-formation ability.
Glass Transition Temperature
Conventionally defined as a temperature at which shear viscosity becomes equal to 10^{13} P. When a liquid alloy is rapidly cooled, its viscosity increases. At some point, the atomic mobility becomes too small to keep the liquid in internal equilibrium, and soon melt becomes homogeneously frozen. This temperature is called "glass transition temperature". Since the liquid appears frozen when the viscosity is about 10^{13} P, this definition was adopted. This structural freezing to the glassy state depends on the cooling rate, so the glass temperature cannot be uniquely defined for a specific composition of alloy.
Johnson-Mehl-Avrami Plot
Solid-state nucleation and growth are described by the Johnson-Mehl-Avrami equation:

$$\alpha(t) = 1 - \exp[-b(t - t_0)^n],$$

where $\alpha(t)$ is the fraction transformed after time t, t_0 is the incubation time, b is a rate constant, and n is an exponent dependent on the type of transformation. From the plot of the double logarithm of $1 - \alpha(t)$ as a function of $t - t_0$, one can determine the exponent n.
Laser Glazing
A processing technique in which a high-power, pulsed laser is used. The laser pulse can be used to rapidly melt the surface layer of a crystalline solid. Heat from the melted layer is then rapidly removed through conduction to the bulk achieving very high cooling rates. In laser ablation, pulses of light are used to vaporize the surface layer of a material. The vapor is then deposited on a substrate and can form an amorphous solid.
Mechanical Alloying
A process to produce alloy by ball milling. Powders of different elements or alloys are mechanically mixed through high-energy ball milling. While milling, new phases are formed. Many amorphous or nanocrystalline alloys can be obtained by this method.
Melt Spinning
A technological process in which a molten alloy is ejected onto the surface of a fast rotating drum. It allows to achieve high cooling rates (10^5–10^6 K/s). Frequently used to obtain metallic glasses.
Metallic Glass
Glass that exhibits metallic characteristics, such as electrical conductivity,

reflectivity etc. Formed in metal–metalloid and metal–metal alloy systems, usually by rapid quenching from the melt.

Microcrystalline Model
An abandoned model of the structure of metallic glass suggesting that metallic glass is a conglomerate of microcrystals.

Nanocrystallization
A process that leads to the formation of a material with grains of the size order of 10 nm and up. Examples are mechanical alloying and rapid crystallization of a metallic glass.

Pair Correlation Functions
Functions describing correlations of atomic pairs in space. See atomic pair distribution function.

Plasma Spraying
See spray processing techniques.

Polymorphous Crystallization
A crystallization process in which the composition of a crystallizing phase is the same as the glass matrix. This does not require diffusion.

Primary Crystallization
A crystallization process in which one phase is growing with a different composition than a matrix.

Quasicrystal
A recently discovered special type of an ordered phase with an incommensurate structure characterized by an irrational ratio of lengths such as the golden mean $(1 + \sqrt{5})/2$. A quasicrystal has the point-group symmetry of an icosahedron. It was believed for a long time that materials exhibiting icosahedral symmetry could not exist.

Rapid Quenching Process
A process in which heat is removed from a material at a very high rate. This process is used, for example, to preserve phases that are only stable at high temperatures, at room temperature, and to create metastable phases.

Rapid Solidification Process
Various technological processes that transform liquid or gas phases into a solid state on a very short time scale, thus not allowing for significant structural atomic rearrangements and usually favoring the formation of a metastable crystalline or amorphous solid.

Rapid Thermal Annealing
A heat treatment process in which the temperature of a sample is raised rapidly with a rate of about 10^2–10^4 K/s. These rates can be achieved using a pulsed laser or by passing pulses of current through the sample.

Solid-State Amorphization
A process in which the transformation to an amorphous phase is achieved in the solid state, without going through a liquid state. An example is radiation damage during which a number of structural defects occur and eventually the crystalline lattice disintegrates to form a glass. Another possibility is diffusive reaction induced by low-temperature annealing or mechanical alloying.

Spray Processing Techniques
Processing techniques using dispersed, atomized droplets which are impacted against a substrate. The alloy can be melted and then atomized, or a fine crystalline powder can be injected into a hot plasma flame. The plasma flame will melt, accelerate, and project the molten droplets onto a substrate.

Sputtering
A process in which a high-energy argon plasma is used to kick out atoms from the target. These atoms then settle down on a substrate forming a film. A low bias field is applied to reduce the energy of the sputtered atoms and avoid resputtering, or damage to the film on the substrate.

Structural Relaxation
Subtle rearrangement of atoms in the material upon annealing resulting in a decrease of the free energy. Many physical properties sensitive to local atomic structure, for example atomic diffusivity, viscosity, ductility, magnetic anisotropy, are affected by the structural relaxation. It should be noted that, while the overall change in physical density during structural relaxation is very small, $\approx 0.5\%$, changes in some other properties can be much larger.

Structure Factor
An experimentally observed quantity in X-ray, neutron, or electron diffraction. Sometimes called the interference function. The structure factor is a Fourier transformation of the atomic pair distribution function. In crystalline solids, structure factors describe the scattering of atoms in the unit cell. The interference function is due to three-dimensional arrangements of the unit cells.

T-T-T Curve
The Time–Temperature-Transformation curve. Describes time as a function of reduced temperature T/T_l (T_l is the liquidus temperature) to transform a small portion of undercooled liquids into a crystal.

Undercooled Liquid
A liquid phase at a temperature below the melting point under normal pressure. Undercooled liquids can be obtained by rapid quenching techniques. During quenching from the melt, the first liquid phase becomes undercooled, and then freezes at the glass transition temperature into a glassy phase.

Vapor Deposition
A technological process in which vapors of different elements are deposited on a cold substrate.

Vitrification
A process leading to the formation of a glass.

Subject Index

Al 7075-T6 6, 9, 11, 27
Al-12 wt.% Mn 6, 9
Al-14 wt.% Ni-12.6 wt.% Mm 6, 11, 13
Al–Be System 19, 75
Al–Cu–Fe 9, 45
Al–Fe–Ce 4, 44
Al–Mm3 29
Al–Si 8, 29
Al_3Ni 32
Al_3Zr 8, 113
Al_6Mn 23
aluminum flakes 30, 31, 32, 36
aluminum refractory 32, 197
amorphous beryllium 13, 24, 155
anelastic 62, 63, 67
artificial state 4, 5
atomic distribution functions 12, 49, 50, 107
atomic structure 12, 60, 63, 68, 107

behavior 3, 4, 11, 135
beta phase 6, 71
binary amorphous alloys 6, 76
Bragg peak 11
brazing alloys 76
bulk 6

C curve 8, 9
catalyst 11
cellular growth 12
characterization 49
chemical bonds 9, 11, 13, 172
chemical properties 177, 178
chilled block 9, 36
chlorofluorocarbon 214
Co concentration 9
Co_4–Fe_5–B_{18}–Si_3 9, 103
coarsening 23
coercivity 11, 165
cold rolling 137, 138
compaction 29
component 6, 14
composition 24

composition, short-range order 11, 13, 28
compression 42
consolidation techniques 42
cooling rate 2, 11, 13
coordination number 2, 13, 67
CoP 103, 146, 151
corrosion-resistant alloys 187, 188, 190
creep 143, 146, 148, 149, 150
creep deformation 146
creep rupture 146
crystalline alloys 10, 11
crystallization 11, 13
curie temperature 157

decomposition 11, 13, 16
dendritic structure 7, 13
dense random packing 49, 50
die 36
differential scanning calorimetry 51, 52, 55
diffraction 11, 72
diffraction methods 53, 72
diffraction patterns 54
diffraction theory 11
diffusion 16, 93, 96, 98, 106, 110, 130
dimensionless variable 110
disorder media 13, 68, 69
distribution transformer 70, 72, 73
domains 163, 164, 168
droplet method 29
droplet size 29, 30
ductility 137, 142

effective-medium approximation 106, 107, 110, 112
elastic 63, 67, 135
electrical resistivity 61, 177
electrical transport 61, 64, 172, 177
electrodes 202
electrodischarge 202, 204, 205
equation of diffusion 16
eutectic crystallization 3, 9
experimental diffraction techniques 11, 72, 73

explicit equation 106
extended solid solubility 9
extrusion 137

failure 135, 136, 137, 138, 221
fatigue 142, 143, 151
Fe–Cr metalloid 9
Fe_{40}–Ni_{40}–P_{14}–B_6 9
Fe_{80}–B_{20} 9, 10, 176
feedstock 9, 13
five-fold symmetry 12
force 9, 10, 13, 127
Fourier analysis 8, 9, 11, 13, 61, 62
fracture toughness 42, 137
free-jet-chill-block 3
freeze 4

gas atomization 8
glass 2, 4, 8, 175
glass-forming composition 8, 9, 10
glass formation 8, 175
glass-formation criteria 11, 13
grain 61, 135, 137, 146, 212
grindability 218, 222
growth 4

Hall effect 61
Hall petch 61
heat sink 26
heat transfer coefficient 26
high cooling rate 11, 26
high corrosion resistance 187, 188, 191
high performance 187, 189
high-resolution electron microscopy 85, 88, 89
high-speed tool steels 187, 188
homogeneity 224, 225
hot deformation 146, 187
hot rolling 187
hydrogen isotope 115

incipient crystallization 11
induced anisotropy 118, 167, 169
infrared thermography 209, 214
instabilities 11, 13
integral breath 11, 13
internal friction 123, 124, 125
ion beam 127, 130, 131
isolated 30, 31, 35

Johnson-Mehl-Avrami plot 59, 60, 239
joining 218, 229, 230

kinetics of solidification 34

laminar flow 36
laser 118, 239
laser glazing 118, 239
liquid dynamic compaction 6, 13
liquid-to-crystal transition 4
liquidus 3, 13
lithium 112

magnetic after effect 118
magnetic anisotropy 167, 169
magnetic hardening 166
magnetic moments 160, 161, 163, 227
magnetic properties 169, 180, 226
magnetic recording 163, 225
magnetic short-range order 11, 13, 67
magneto-optical 182, 184
magnetostriction 161, 182, 184
mass-production 184
mechanical alloying 239
mechanical properties 153, 218
melt extraction 3, 4, 23, 41, 239
melt spinning 3, 4, 23, 35, 36
melt zone 4, 6
metallic glasses 4, 6
metallographic 35, 45
metastable crystalline phases 7, 13
metastable states 13, 17
Mg–Zn–Al 2, 4
microhardness 61, 138, 142
microsectioning 55
microstructure 60, 61
modelling 30, 31, 110
morphology 23
Mössbauer spectroscopy 60, 61

nanosecond 156
Nd–Fe–B 9, 43
Newtonian viscosity 9, 11, 28
Ni–Fe–Mo–B 91
Ni_3Al 8, 9, 77
non-equilibrium 4, 9, 11, 13
nucleation 4, 10, 11

p–d bonding 23, 42, 46, 169, 170, 171, 173, 174
packing density 40
pair correlation function 12, 13, 23
pair distribution function 12, 13
particle size 30, 46
particulate 30, 46
permanent magnet 157, 179
permeability 158, 159
phase diagram 6, 17
phase separation 6, 11, 13, 55
phase transformation 13

physical phenomena 12
picosecond 4, 10, 11, 16
planar flow casting 135, 139
plasma spraying 42, 127, 130
plastic flow 142, 143
Pol Duwez 2, 7, 14, 23
polycrystalline 11
polymer bonding 187, 191
pores 9
porosity 9, 11
position 11, 50, 80
positioner 11, 50, 68, 80
positive temperature dependence coefficient 13
powder pattern 29, 30
power magnetic application 179, 182, 227, 229
precipitation 24
primary recrystallization 1, 4, 6, 8

quasi-equilibrium 23
quasicrystalline 23
quenched 11

radiation sources 50, 51, 53, 72, 96
radiative heat transport 72, 96, 240
radiotracer 3, 7
recrystallization 11, 240
recycling 188, 210
reduced atomic distribution functions 49, 50
relaxation 93, 96, 106, 241
rotating water bath 37

scattered intensity 24, 62, 63, 67
selected area diffraction 24
self-diffusion 98, 100, 102
sensors 182, 183
sheer 135, 137
shielding 12, 13
short-range order 11, 13, 28, 67
size refinement 30, 218
slab 2, 3
solubility extension 13, 14
specific heat 23
spherodization 23

split d-bands 46
spray deposition 3, 30
spray processing technique 3, 30
sputtering 30, 46, 241
stability 13, 18
strength 138
structural relaxation 49
structure factor 11, 49
structure image 37, 38
surface melting 38
surveillance 182, 183

tension 135, 137, 221
thermal conductivity 61, 63, 68
thermodynamics 4, 6, 8
time–temperature-transformation 2, 4, 5, 7
T_0 line 9, 10
T_0 melting temperature 9, 10
topological 11, 38
transition temperature T_g 9
transmission electron microscopy 24, 49, 52, 55, 84, 85, 86, 88, 89
triaxial 142
tungsten alloys 146, 151

ultrafine grain 137, 146, 221
ultrasonic vibrator 135, 137
undercooling liquid 4, 18, 90, 241
undercooling 4

vanadium alloys 9, 10
vapor deposition 30
viscosity measurement 10, 11, 16, 124, 125
vitrification 11

wear 142, 143
wear resistance 142, 146, 151

X-ray radial distribution function 11, 50, 68, 80

yield 135, 137, 146

Zr–Cu 9, 10

Springer Series in *Materials Science*

Advisors: M. S. Dresselhaus · H. Kamimura · K. A. Müller
Editors: U. Gonser · R. M. Osgood, Jr. · M. B. Panish · H. Sakaki
Managing Editor: H. K. V. Lotsch

1 **Chemical Processing with Lasers***
By D. Bäuerle

2 **Laser-Beam Interactions with Materials**
Physical Principles and Applications
By M. von Allmen and A. Blatter 2nd Edition

3 **Laser Processing of Thin Films and Microstructures**
Oxidation, Deposition and Etching of Insulators
By. I. W. Boyd

4 **Microclusters**
Editors: S. Sugano, Y. Nishina, and S. Ohnishi

5 **Graphite Fibers and Filaments**
By M. S. Dresselhaus, G. Dresselhaus, K. Sugihara, I. L. Spain, and H. A. Goldberg

6 **Elemental and Molecular Clusters**
Editors: G. Benedek, T. P. Martin, and G. Pacchioni

7 **Molecular Beam Epitaxy**
Fundamentals and Current Status
By M. A. Herman and H. Sitter 2nd Edition

8 **Physical Chemistry of, in and on Silicon**
By G. F. Cerofolini and L. Meda

9 **Tritium and Helium-3 in Metals**
By R. Lässer

10 **Computer Simulation of Ion-Solid Interactions**
By W. Eckstein

11 **Mechanisms of High Temperature Superconductivity**
Editors: H. Kamimura and A. Oshiyama

12 **Dislocation Dynamics and Plasticity**
By T. Suzuki, S. Takeuchi, and H. Yoshinaga

13 **Semiconductor Silicon**
Materials Science and Technology
Editors: G. Harbeke and M. J. Schulz

14 **Graphite Intercalation Compounds I**
Structure and Dynamics
Editors: H. Zabel and S. A. Solin

15 **Crystal Chemistry of High-T_c Superconducting Copper Oxides**
By B. Raveau, C. Michel, M. Hervieu, and D. Groult

16 **Hydrogen in Semiconductors**
By S. J. Pearton, M. Stavola, and J. W. Corbett

17 **Ordering at Surfaces and Interfaces**
Editors: A. Yoshimori, T. Shinjo, and H. Watanabe

18 **Graphite Intercalation Compounds II**
Editors: S. A. Solin and H. Zabel

19 **Laser-Assisted Microtechnology**
By S. M. Metev and V. P. Veiko

20 **Microcluster Physics**
By S. Sugano

21 **The Metal-Hydrogen System**
By Y. Fukai

22 **Ion Implantation in Diamond, Graphite and Related Materials**
By M. S. Dresselhaus and R. Kalish

23 **The Real Structure of High-T_c Superconductors**
Editor: V. Sh. Shekhtman

24 **Metal Impurities in Silicon-Device Fabrication**
By K. Graff

25 **Optical Properties of Metal Clusters**
By U. Kreibig and M. Vollmer

26 **Gas Source Molecular Beam Epitaxy**
Growth and Properties of Phosphorus Containing III-V Heterostructures
By M. B. Panish and H. Temkin

---

* The 2nd edition is available as a textbook with the title: *Laser Processing and Chemistry*

Springer and the environment

At Springer we firmly believe that an international science publisher has a special obligation to the environment, and our corporate policies consistently reflect this conviction.

We also expect our business partners – paper mills, printers, packaging manufacturers, etc. – to commit themselves to using materials and production processes that do not harm the environment. The paper in this book is made from low- or no-chlorine pulp and is acid free, in conformance with international standards for paper permanency.

Printing: Saladruck, Berlin
Binding: Buchbinderei Lüderitz & Bauer, Berlin